GREGG BRADEN

Awakening to Zero Point

The Collective Initiation

Definition of Zero Point

The amount of vibrational energy
associated with matter, as the parameters
defining that matter decline to zero.

To an *observer,* the world at Zero Point appears
to be very still, while the *participant* experiences a
quantum restructuring of the very boundaries
that define the experience.

Earth and our bodies are preparing for the
Zero Point experience of change,
collectively known by the ancients as
The Shift of the Ages.

GREGG BRADEN

Awakening to Zero Point
The Collective Initiation

Ancient prophecies predicted it. Indigenous traditions honor it. Changes within the Earth are affecting your sleep patterns, relationships, the ability to regulate your immune system and your perception of time. You are living a process of initiation that was demonstrated over 2,000 years ago, preparing you to accept tremendous change within your body. That change is happening now.

The Shift of the Ages has already begun!

PUBLISHED BY
RADIO BOOKSTORE PRESS
BELLEVUE, WASHINGTON, U.S.A.
A Sacred Spaces/Ancient Wisdom Book

Published By:
Radio Bookstore Press
P.O. Box 3010
Bellevue, WA 98009-3010
www.radiobookstore.com

Radio Bookstore Press books and related information may be purchased for educational, business, sales or promotional use. For information and a free catalog, please write:
Radio Bookstore Press, P.O. Box 3010, Bellevue, Wa. 98009-3010.

Cover Concept: Gregg Braden
Cover Art: Joel Whitehead, Taos, New Mexico
Book Design and Composition: Webb Design Inc., Taos, New Mexico

REVISED EDITION

9 8 7 6 5 4 3 2 1 • 96 97 98 99 00 01 02

ISBN 1-889071-09-9
Printed in the United States of America

DEDICATION

This work is dedicated, with the highest love
and in the greatest truth,
to you who have journeyed so far from home,
knowing unspeakable joy,
enduring unbearable pain,
allowing the possibility of our Earth experience.
Your integrity has remained impeccable,
your companionship an honor.

AW RA
The Resurrection Of The Light

VEY ATA TIM SHAL BOW
An Opening Has Occurred Into An Inexhaustible
Field Of Knowledge And Meditation

TOW VA AW RA
The Light Of Good Is Once Again

(Adapted from the initiation "codes"
of an ancient Hebrew text)

TABLE OF CONTENTS

ILLUSTRATIONS

CHAPTER 1

CHAPTER 2

CHAPTER 3

CHAPTER 4

LIST OF TABLES

ACKNOWLEDGMENTS

It is not possible to reference, by name, all who have contributed to the creation of *Awakening to Zero Point: The Collective Initiation.* Each individual whose life path has become part of mine has contributed in some way to this material; each was my teacher during the time that we shared. To those who have believed in this work and waited patiently following the mysterious disappearance of the original manuscript in 1991, my deepest gratitude. Your support, honesty and lives are reflected in various aspects of this *Awakening to Zero Point.*

My sincerest thanks to:

Paul and Laura Lee for your friendship and support. Laura, a special thank you for your insight and coining the term *Collective Initiation.*

Drunvalo Melchizedek for graciously and generously sharing in the *Flower of Life*, the science of *Mer-Ka-Ba*, and your contributions to this work. Most importantly, thank you for your friendship.

Sananda Ra for your music, sense of adventure and friendship. From you I have learned how to travel in third world countries and lead others, with confidence, to the Holy and Sacred Spaces of the Earth.

Sharon and Duaine Warren, and the Arizona Chapter of the Center for Crop Circle Studies (CCCS) for your honesty and support. Your time and willingness to contribute has been invaluable to me, your friendship an honor.

Gil and Carol Gray, for your support, contributions and encouragement during the "challenging years." Our time together, as well as your friendship, is cherished deeply.

Sylvia Braden, my mother, for your patience with me during my formative years and your openness to new ideas. Your love is appreciated more than I have ever expressed.

Additionally, I would like to thank Alex Gray for his willingness to share his "life search" through the visionary artwork of Sacred Mirrors. Alex, your generosity in allowing me to share that work with others has helped to convey the aliveness of the human experience. Dan Winter, thank you for graciously allowing me to share your work from *Alphabet of the Heart* to validate the resonant relationship between the Earth, heart, brain, cell and immune systems. Your work brings a tangible link to the inner knowing for all whose lives you touch.

A Prayer of Clarity

I give thanks for the opportunity to join together as one, in mind and heart, throughout the time that you are engaged in *Awakening to Zero Point: The Collective Initiation.* It was you in the eye of my mind as this material coalesced. It is to you that I dedicate all of the love and understanding that may result from this text, for it is you who are ushering in a new wisdom.

Each time one individual remembers, the bridge to that memory becomes more accessible to the next. You are both pioneer and midwife. Through you we push the boundaries of human experience to a place where no one has experienced before. Through you we birth a new world in the process.

From you I invite an open heart and an open mind as this material unfolds before you. I ask patience and tolerance for my words if they are not your words. Please read with your heart as well as your mind, feeling the intent underlying each word and each thought.

For myself, I have asked for the wisdom to find the words to convey the Awakening material in a format that is meaningful to you. May my words address your concerns and answer your questions, even before they are asked.

For each of us, I ask the wisdom to recognize truth, and the strength to live that truth, with integrity and love, in our daily lives. In the presence of the infinite creative intelligence that lives within.

FOREWORD

THE ZERO POINT PHENOMENON

BY LAURA LEE

Do you feel like time is "speeding up?" Are you experiencing changes in your sleep patterns and dream states, alternating periods of "black hole" sleep with periods of intense and vivid dream activity? Have your emotions and relationships intensified? Is *déjà vu* a common experience for you? What about the vague feeling that something is different now, that somehow, you've been through this before? You may be happy to discover, as I was, that you are not just imagining it. You are not alone. And there are very real reasons for these experiences.

Evidence is accumulating from diverse sources to suggest that these experiences, psychological though they may be, have a physiological and geophysical component. This means that changes within Earth's body affects our bodies, because the two are tied in subtle ways.

So subtle, in fact, that we are mostly unaware of this "bonding" until viewed in the life or fertility cycles of many species that are timed to the tides or full moon, or in the recently discovered magnetite, a specialized brain cell, found in abundance in the brains of mammals (including humans) and birds that tune and respond to Earth's magnetic field. Allowing homing pigeons to home in, sea creatures to navigate migration patterns, magnetite may also be the key to understanding why some people and animals are sensitive to earthquakes before they happen (they sense the localized anomalies in the magnetic field known to occur in the hours and days prior to seismic events) and perhaps why sheep have been seen to sit in rows upon ley lines, natural lines of magnetic force in the Earth. So, we are tuned and we are affected. What are the changes within the Earth affecting us?

They are cyclical, occurring in extremely long time waves, by human standards. That's one of the reasons we are mostly unaware of them. The best place to view the clues left behind by previous cycles is in geological record, the book kept by Earth herself. Science knows little about how to read these cycles, or what they mean. That information may be contained in the records of the previous cultures who lived through them.

Breakthrough research comes from Gregg Braden, who has correlated the two records and pieced together a fascinating, and, if he's right, important picture. Braden's career history and personal experiences had much to do with recognizing this new evidence. Former experience as a computer systems designer and geologist led him to recognize the evidence of these geophysical cycles in the geologic record. Two Near-Death Experiences at an early age, and years spent guiding tours to sacred sites throughout the world, led him to research the temples, texts, myths, and traditions of various ancient cultures.

Braden found that previous cultures had not only experienced and left records of their experience of these cycles, they found them useful for easing access to higher states of consciousness. So useful, that in between cycles, they designed and built temples, or utilized natural sites, that exhibit these same geophysical, cyclical conditions, the same ones that are in exponential transition on Earth today. They called this point in the cycle "The Shift of the Ages." What's more, they left us the instructions.

GEOPHYSICAL CONDITION #1
EARTH'S FLUCTUATING BASE FREQUENCY

Earth's background base frequency, can be likened to a pulse or heartbeat. Though the studies are conflicting and controversial, there is evidence that this pulse rate is fluctuating. The mechanics of what drives the pulse rate, and how it affects us, are unknown. Braden suggests that this fluctuation is significant, in that Earth's vibration rate affects our own.

GEOPHYSICAL CONDITION #2
EARTH'S DIMINISHING MAGNETIC FIELD

It is agreed that the Earth's magnetic field strength is declining. According to Professor Bannerjee of the University of New Mexico, the field has lost up to half its intensity in the last 4,000 years. And because a forerunner of magnetic polar reversals is this field strength some believe that another reversal is due. Braden believes these cyclical Shifts are associated with reversals, so Earth's geological record indicating magnetic reversals also marks previous Shifts in history. Within the enormous time scale represented, there were quite a few of them.

THE BOOK OF EARTH PAST

The geological record is like one big book whose sedimentary pages record the events of their day. Magnetic pole reversals leave their mark in the sea floor's spreading ridges, where the once molten rock's iron particles aligned to the north pole as the lava cooled and hardened. Today, through core samples, we can "read" that the magnetic orientation to North shows periodic 180 degree flipping. Over a period of 76 million years, 171 reversals are recorded; nine of them in past 4 million years. "My suspicion is that the magnetic flip-flop occurs very quickly once Earth's magnetic field has diminished near the Zero Point, then slowly builds up again," says Braden.

"It may happen soon, or thousands of years from now—it's impossible to pinpoint exact dates in the geological record," says Vince Migliore, editor of *Geo-Monitor*

Newsletter. "But we know that Earth's geomagnetic field, known to fluctuate in it's intensity in the recent geologic past, is now just a fraction of what it has been historically." What is it right now, on a scale of 10, 5 being average, I asked him. The answer: 1.5! What do scientists know of the effects? According to Migliore, a common phenomenon is tremendous migration of the magnetic poles. There are reports of such magnetic anomalies, picked up by compasses, ranging up to 15 to 20 degrees away from magnetic North.

TEMPLE MECHANICS

Braden has measured many ancient temples exhibiting unusual magnetic fields and frequencies. Anecdotal reports concur: from the 1800s to the present, people have reported hearing ringing and hums, seeing strange glows, and feeling sparks from megalithic standing stones and the Great Pyramid.

As a resident of New Mexico, Braden spends considerable time among the temples of the Southwest. He points out that what are called temples in other countries are called merely "Indian ruins" in this, and that influences the way we think about and care for these legacies. In the circular underground kivas built by the Anasazi culture of a thousand years ago, Braden sees "tuned resonant cavities" for the purpose of eliciting various altered states of consciousness. A resonant cavity is a hollow space, the dimensions of which have a naturally occurring frequency that sets up a resonance, or harmonic feedback loop, and tunes with another frequency. "In the case of certain kivas," says Braden, "that other frequency is the human mind."

VISIT TO A KIVA

He tells the story of a tour of one well preserved kiva, where his group was told they couldn't play Native American cedar flutes as they had in the past. "I was looking forward to experiencing the kiva's unusual acoustics and the meditative reverie that often results from it. But a park ranger met us, saying that musical instruments were now prohibited, because a few days prior, a visitor had died of a brain aneurysm while blowing a conch shell in the middle of this kiva. The park wasn't willing to risk any more accidents."

My husband Paul had a much happier experience in that very same kiva. With Braden as tour guide, Paul and I spent several days hiking dusty trails linking kivas of different dimensions (and frequencies) to outposts and rock outcroppings, some decorated with petroglyphs. Walking through a kiva, Paul stopped and stood, listening as with his inner senses. He sat down in the middle, closed his eyes, and tuned in. It was so powerful, with such intense emotional energy, he later reported, that he moved to the kiva's edge, sat on a ledge, continued his meditation, and there had an awakening that he feels changed the course of his life.

INNER MECHANICS

Is it any accident that many temples and sites known as sacred offer a localized experience altered mechanics and frequencies? Some studies reveal that the upper known chambers of the Great Pyramid have significantly lower magnetic readings than the lower chambers, while significantly higher than normal frequencies have been measured. Braden believes that the Great Pyramid is just one example of a technology developed by previous cultures to recreate these conditions between cycles, in chambers especially designed for initiation into altered states of consciousness.

Just what effect does a lesser magnetic field and higher vibratory rate have on us?

"The opportunity to more easily change the patterns that can determine how and why we love, fear, judge, feel, need, and hurt." says Braden. "Dense magnetics lock in emotional and mental patterns from generation to generation in the morphogenetic field. With lesser magnetic fields, this seems to ease up, allowing easier access to higher states, as the cells of our body tune to, and try to match, Earth's frequency like tuning forks, thereby altering our own."

Records of Previous Shifts

Legacies both written and oral indicate that ancient cultures world wide have experienced the "Shifts" of previous cycles. "The ancient record keepers left us the markers for this event, what to expect, and, very importantly, a strategy for these times," says Braden. "The strategy involves making the most of the opportunity for access to higher states of consciousness, and focuses on the importance of emotions. The Essenes made a science of this. And the ancient Egyptian mystery school initiations took place in specific temples in sequence, one before the other. Each temple, and personifying temple deity, was dedicated to one aspect of the human psyche. Each aspect must be balanced prior to moving to the next level."

Initiation Rites

Why the focus on emotions? Looking through current biotech research, Braden found that our very DNA, our life codes, are affected by our emotions. The way he explains it, DNA coding options come up as "electives" at various times, able to turn on or off. Emotions trigger specific biochemicals, which influence the chemical voltage and frequency of cells, to which molecules such as DNA respond. Therefore, it is possible for our emotions to act as "switches" for "turning on" options within our DNA to make new amino acids (reports of spontaneous mutations are on the increase) in preparation for an evolutionary leap. Ancient wisdom, then, may offer a successful strategy for life today, one that works on many levels.

Earth as Global Temple for Collective Initiation

"We are living the completion of a cycle that began nearly 200,000 years ago, and a process of initiation that was demonstrated over 2,000 years ago," says Braden. "In past ages, through proper initiation, these special conditions were utilized for clearer access to higher states. Now we don't have to recreate them in specialized temple chambers. We don't have to go anywhere. We are living in a global initiation chamber, with these geophysical conditions occurring on a world wide scale. It's as if Earth herself is preparing us for the next stage of evolution."

Science may be witnessing events for which there are few points of comparison, but, says Braden, "ancient traditions have preserved the understanding that during key moments in human history, a wisdom has been offered allowing individuals to experience rapid change without fear. This is one of those moments. This wisdom is now being passed down. Your life is preparing you for the Shift. The recorded and predicted timetable is intact. The time is now."

The Chinese have a saying, "May you live in interesting times." For those frightened of change, who equate dull with secure, this is meant as a curse. Yet interesting times, if today is any indication, may also be a rare window of opportunity, a chance to get it right this time around, a collective journey, and certainly, High Adventure. I find that it is a personal choice.

Laura Lee contributes to many national magazines, and is host of the nationally syndicated talk show on radio, The Laura Lee Show.

Introduction

When you once see something as false
which you have accepted as true,
as natural, as human,
then you can never go back to it.

ADAPTED FROM J. KRISHNAMURTI

It was late afternoon, January, 1987. Though I was in Egypt, it was winter, and the cold was becoming apparent as the sun set behind the peaks that towered above me. The previous evening temperatures had dropped into the midteens and there were still traces of snow in the shadows where the sun had not reached. I had found a rock ledge jutting out from the cliffs to my back and I looked up toward approximately two thousand crude stairs of stone winding to the peak of this most holy mountain of Moses, Mt. Sinai. Built and maintained by monks in the monastery three thousand feet below, many of the steps were still covered with slush that had started to glaze over as the evening winds picked up. Our group had left the village of St. Catherine nearly nine hours earlier, some on foot, some by camel. The others had already started back, looking for an alternate route that would cut their hiking time in half.

It felt good to rest my legs as I stared up toward the rough stone steps. Regardless of which route I chose, I would be hiking down in the dark. I was in no hurry to leave.

Though not surprised, I was still in awe of the cumulative set of circumstances that had led to this sacred journey, allowing me to witness and experience dusk on Moses' mountain. The first near death experience of electrocution early in life that, quite literally, had shocked me into an awareness of other realities. Within weeks, the second near death experience of drowning, reminding me of what it felt like to "let go" and allow. Through each of the experiences I had been reminded that through shifting feelings within my body, I could shift my perceptions of the outside world. The resulting high fevers that I experienced as an adult, sometimes lasting for days at a time. Just nine weeks earlier I had experienced one continuous fever that raged between 104 and 106 for seven full days. Immediately afterward, I found myself in Taos, New Mexico and part of an amazing series of events that led me to this moment.

I don't know when it began. I did not notice the shift in my awareness until my mind was flooded with the emotion of two questions forming a singular thought. I found that thought overwhelming me, blocking any other idea or feeling that may have been present at the time. The questions were familiar. I had asked them of myself many times before. Always before I was left without resolution. That night on Mt. Sinai, something changed. The questions were right there.

The first question:
Of all of the things that I may do with my life, within any given moment, what is the greatest gift that I may ever offer to myself, or to another?

The second question:
If I were to leave this world tomorrow, and look back upon my life, and all that I have accomplished until this moment, would I feel complete?

Relief filled my body as the answers became clear. I climbed down from the ledge and found the trail that would lead me to the monastery and eventually back to the village of St. Catherine. I was alone in thought for the duration of the three-hour return and arrived in the village at approximately 8:30 PM.

Following dinner I found myself outdoors in prayer and meditation. I sat on the dry, sandy soil wrapped in a prayer blanket behind the stone bungalow that was home for the night. It was a beautifully clear, cold, and moonless night in the high Sinai desert and I remember thinking of how similar it felt to the mountain nights of northern New Mexico. Still in prayer, I began to notice a ringing in one ear and then the other (as of this writing, the ringing continues), followed by a rapid increase in my heart rate and respiration. Something was happening! I opened my eyes and, for some reason, looked to the sky directly above me. Immediately I noticed what appeared to be a mist forming over my bungalow, as well as the adjoining unit sharing a common wall. I stood up and walked a few yards away to get a better view and noticed that the sky was still perfectly clear with the dense clusters of stars forming the edge of the Milky Way especially well defined. The cloud hovering over my unit became more dense as I returned to the porch and my meditation. Upon closing my eyes, a very uncomfortable sensation filled my body. I opened my eyes to see the mist fading from sight.

Seconds later the same process began again, only this time the feeling was very different. Still in meditation, a tremendous feeling of vastness filled my body. The mist grew more dense as it settled to the ground, completely engulfing myself and both units of the stone building! (The following morning, my neighbors in the adjoining room remembered nothing of the previous evening.)

As I became comfortable with the sensations, I realized that this was an opportunity that I had always suspected would present itself, the opportunity to remember. I also knew that the wisdom from that night would come only through living each day whatever knowledge I was offered in those brief moments. I began asking the questions that had never been answered. My body vibrated into a rapid, rhythmic pulse. My heartbeat became a flutter. In the cool desert air, my clothing became wet with the perspiration generated from the experience.

Quickly I learned that my asking was getting in the way, as the answers came before the questions were completed. I stopped asking and began to listen as a flood of information cascaded through my body continuously for what seemed like forever and an instant, simultaneously. In that space, time had no meaning. I later determined that the entire experience had lasted approximately 22 minutes.

During my time of meditation, I found unobstructed access to even the most minute detail of any information that my mind could conceive of. I began to see the information from a perspective that mirrored my lifelong belief in the unity of all things. It was during that time that a voice, a familiar voice from my near-death experiences, asked me to remember the feeling that I was experiencing in those moments. Additionally, I was reminded that access into this field of knowledge was accomplished through creating that very feeling within my body. To remember I simply was asked to feel!

That night I made a conscious choice in my life that would forever change the way in which I related to my wife, my family, my friends, career and to every individual that would ever come into my life again. I knew that to occasionally create the feeling of that night, I would never be fulfilled in my life. My choice of mastery would not result from merely thinking the feeling in my mind. That night I chose to *become* the feeling that was offered to me, as a point of reference, in this desert prayer.

In the cold Sinai desert that night in 1987, I experienced my first group *Mer-Ka-Ba*; the Light Body Spirit vehicle of a many and singular consciousness. What I gained from the evening on Mt. Sinai, and subsequent journeys to the Andes Mountains of Bolivia and Southern Peru, were the tools to offer the Awakening information in a concise and meaningful format, relevant to our lives today.

What is the greatest gift that I may offer you? I believe the answer is simply the gift of yourself, in wholeness, in the absence of conditioned patterns that may no longer serve you. My offerings of books, multimedia workshops, seminars and sacred journeys are designed as a stepping-stone toward that end. In part, these offerings also serve to answer the second question.

If I were to leave this world tomorrow, would I feel complete?

Having offered the *Awakening to Zero Point* information over these last few years, my answer to that question would be, yes... almost...

My intent in offering this material now is to provide a concise and accurate synthesis of the events unfolding within our solar system, within our world and within each living cell of our bodies, as all are related. It is my goal, and sincere hope, that the information within these pages will provide continuity to our understanding of planetary, human and personal evolution.

Our lives, our day-to-day experiences of pleasure, pain and all that we have ever felt, touched, created or uncreated (nothing is ever destroyed) have occurred as components of an ongoing process. This process is preparing us to accept and gracefully accommodate tremendous change within our world and our bodies. The processes unfolding are about us, of us and for us. Through the offering of the *Awakening* material, a vocabulary may be remembered and continuity developed within which we may express outwardly that which we are feeling within. In the outward expression, our memory becomes anchored even more firmly into the world, becoming even more accessible to others.

Awakening to Zero Point is intended to offer a context within which to view the phenomenon of our lives and this time in history. I believe that we have outgrown the need to compartmentalize our knowing world into artificial boundaries of "religion," "metaphysics" or "science." These terms are examples of a language that has attempted to categorize our experience into discrete packages of knowledge. Our journey into the quantum world of the atom, human genetics, the exploration of Earth and space has demonstrated that our imposed boundaries are temporal in nature. In creation there is no geology, physics, biology and chemistry. There are simply patterns of energy within patterns of energy, each expressing as observable systems. Our language and beliefs have divided into many that which is one, contributing to our illusion of separation. There can be no separation in that which is one. Perhaps more accurately, all must be embraced as a portion of the whole in a world of unity. Our delineation of these patterns has served us well, bringing us to the point where we see that there are no boundaries in the study of our world. Beyond persuading, convincing or redefining any existing belief system, this text is designed to simply remind you of possibilities.

Though *Awakening to Zero Point* transcends religion, science and metaphysics you may see how religion, metaphysics and the sciences developed surrounding a core body of information that has been distorted through translations and interpretations. You may ask, if this text is not scientific, metaphysical or religious in its nature, then what is it? I invite you to receive *Awakening to Zero Point: The Collective Initiation* as information, without category or label, representing all possibilities without the boundaries of what "should be."

We may ask others out of choice, and we no longer need to do so out of necessity. All answers relevant to us live within us, as the experiences that we masterfully create as our lives. Our map of truth resides within the living intelligence that breathes as each cell of our bodies, pulsating life force throughout every aspect of our being. We have grown beyond the search that has collectively beckoned to us from our outside world. Now, in this lifetime of completion, we are reminded that our truth lives within. It has been with us throughout every facet of our journey, patiently awaiting the time when we could acknowledge and accept. It is the search that has brought us the lessons, the relationships, the mirror of others so that we could arrive at the point where we are in this moment, back to ourselves.

Almost universally, everyone feels it. Something has changed. Something feels different now, during these days. Everyone feels a shift on some level to some degree. The tension of tumultuous change, poised, positioned, enabled. Some describe it as electric, as a low-level anxiety. Others simply state their feeling that something is about to happen, something big! Some individuals feel it rippling throughout every cell in their body, perceiving that time, and their lives, are speeding up. Others are experiencing a new kind of confusion, as if nothing in their lives really fits any longer. Their world is changing so rapidly that they feel out of control, perhaps unable to keep up. To those having difficulty in "letting go" and flowing with the change, the feelings may be uncomfortable, even painful.

There can be no denying that today in your life, now, is a time of unprecedented change. Regardless of what levels of distinction are used to view world events, the systems that provide the infrastructure to life and society, inclusive of personal systems such as health, finance and relationships, are in a state of dynamic flux. Something is happening!

EVENTS THAT WARRANT CLOSER EXAMINATION

Seemingly on a daily basis, you are witnessing events that ask you to redefine the nature and boundaries of what you believe your world to be. Occurring on a global scale, these events are often discounted as discrete, independent and nonrelated "flukes" and given very little media coverage. When viewed through the eyes of a grand framework of experience, however, the same occurrences provide a degree of continuity to what sometimes appear as random and extreme displays of change. Predicted and expected, many indigenous peoples view this time in history as the completion of a grand cycle whose beginnings remain as a faint glimmer in human memory. Whether you believe in the near-term close of a great super cycle or not, one fact remains. Within a relatively short period of human history, regardless of your age, you have witnessed events that rocked the very foundations of who and what you believe your world is all about. I have compiled a sampling of these occurrences, events that certainly warrant closer examination.

- Thousands of people globally, experiencing a common dream of unprecedented change
  - Almost universally, the vision details tremendous shifts in human consciousness accompanied by equally tremendous shifts within the Earth expressed as social, political, economic, meteorological, seismic upheaval.

- Mass migrations away from large population centers
  - In California alone, estimates indicate that 2.5 to 3 million individuals left the state by the end of 1993. Moving companies such as U-Haul, Ryder, Thrifty and Budget have placed a moratorium on their companies' trucks leaving the state due to the lack of returning vehicles.

- The breakdown of systems that have been perceived as security and survival for many throughout the years, such as the economy
  - "Bailout" plans for record numbers of U.S. banks unable to meet their commitments. Legislation passed in 1991 provides for $70 billion in new funding to the FDIC and the Federal Reserve empowering them to direct cash into failing banks and businesses.[1]

  - Massive cutbacks and layoffs resulting in record numbers unemployed and bankruptcies. In 1992, 1,239 businesses with liabilities of more than $1 million filed for bankruptcy in the first three months of the year compared to 398 for the same time in 1991, according to a Dunn and Bradstreet study.[2]

- Controls on stock market indicators diluting their ability to reflect true market trends. For example, prices are artificially fixed for gold, silver, oil, the value of the dollar and interest rates.

- The end of the Cold War accompanied by a global shift toward democracy (for example, Yugoslavia, the Soviet Union and the socialist states, Bosnia, Haiti, Peru, Chile, etc.)

- Dismantling of the Berlin Wall in 1992, as well as the thinking that originally led to the building of the wall.

- Reduction in global nuclear arsenals. Early 1994 saw the agreement, in principle, of reduction and redirection of stockpiles within the new Russian states, following the breakup of the Soviet Union.

- Unprecedented cutbacks in defense spending and closings of military bases in the U.S. accompanied by the redirecting of wartime technology toward peacetime uses.

• The recognition of evidence indicating that technologically advanced societies *did* exist, then *disappeared* on Earth prior to the melting of the last Ice Age over 10,000 years ago

- Solid archaeological and geological evidence of previous civilizations of high sophistication are asking scholars to rethink the accepted chronology and sequence of human history. For example, the Sphinx is now being dated at least twice as old as previously believed.[3] The technology to build such a structure was not supposed to have existed at that time. Who built it?

- Studies indicating that the exposed blocks now forming the exterior of the Great Pyramid (over 90% of the original exterior was removed in historic times) were not fashioned from naturally occurring limestone. Core samples suggest that these massive blocks may have been *poured in place* to achieve the structural integrity, uniformity and close tolerances averaging 1/1000 of an inch between blocks. Within the last 30 years we have just begun to "learn" this science of creating artificial stone.

- Probable evidence from within our own space program, indicating that technologically advanced societies have existed *on at least one other planet within our own solar system*, with probable connections to ancient sites on Earth. This information has been met with a great reluctance to openly share or discuss the evidence with the public.[4]

- Ongoing attempts to control the publication of ancient wisdom relevant to this time in history. For example, Dead Sea Scroll MMT, one of the most revealing, has been recently copyrighted and as of this writing, still unpublished.[5]

• Dramatic shifts in global patterns of weather
  - Record rainfalls in the U.S. Midwest (1993) resulting in massive flooding along the Mississippi, Ohio and Missouri river valleys and their tributaries. An estimated 50,000 homes were damaged or destroyed, 54,000 people evacuated and at least 50 related deaths.[6]

  - Record rain and melting snow resulted in unprecedented flooding throughout the Pacific Northwest and Sierra Nevadas in late 1995 and early 1996. Early January saw the simultaneous closing of the Phoenix, Albuquerque, Houston, Dallas and Denver airports due to the same massive winter storm.

  - A storm that the National Weather Service has called the *Storm of the Century* covered one third of the nation with snow in 1993 and produced winds up to 110 mph, killing 320 people.

  - 1993 saw measurable snowfall in the normally "snowless" southern states with 13 inches reported in Birmingham, Alabama and 20 inches in Chattanooga, Tennessee.

  - Record high temperatures and related deaths throughout the nation. Phoenix, Arizona recorded 76 days of 100+ degree temperatures in 1993. On at least one day, the Sky Harbor airport in Phoenix was closed due to high temperatures preventing the "lift" that planes require for flight.

  - Record temperatures, well below zero, were recorded in at least 18 major U.S. cities and responsible for at least three deaths over one weekend in early January of 1993. For example, temperatures reached well below zero degrees Fahrenheit in Alpena, MI. (-28), Syracuse, N.Y. (-21), and Rochester, N.Y. (-17).

• NASA studies in 1992 indicate that the breakdown of the ozone layer of atmosphere in the Northern hemisphere could reach 40% within the 1992–1994 time frame, reportedly affecting eyes, skin, immune system, crops and marine life. 1993 witnessed the deepest depletion of Ozone on record.[7]

• Mass experiences of a compelling need for change, looking within to self and alternative modes of healing and understanding to accomplish the change.

This shift is evidenced in the resurgence of the many and diverse offerings, including those of John Bradshaw, Louise Hay, Bernie Segal, Robert Bly, Katrina Raphael, as well as the phenomenon of *The Celestine Prophecy* and *Bringers of the Dawn*, for example.

- The dramatic increase in the occurrence and boldness of reported, and documented, extraterrestrial encounters. These encounters are occurring world-wide, and range from positive experiences with loving and benevolent beings to horrifying experiences of abduction and physical violation, some returning with biological implant devices located under the skin or within the cranial cavity.

- A media campaign helping the public at large become accustomed to the possibility of an Extraterrestrial Presence, as well as, human involvement with that presence. The movies, *Fire in the Sky, Close Encounters of the Third Kind, Cocoon* (I and II), *Starman,* television's *The "X" Files*, *Encounters, Sightings*, Commercials for McDonald's, Budweiser Beer, *Star Trek: The Next Generation* and *Deep Space Nine*, to name a few, as well as an unprecedented number of prime-time specials regarding ancient prophecies and this time in history

- A series of new viruses that have rendered the human immune system defenseless, such as HIV and HIV look-alikes. Neither HIV, nor the new viruses, respond well to conventional treatment. Estimates for 1992 alone predicted that 10 to 12 million individuals would test positive for the viruses.[8]

- New studies released in the fall of 1996 indicating that a portion of the global population had genetically mutated to develop a "resistance" to the HIV and look alike viruses.

- A sudden increase in the number, duration and intensity of seismic events globally
    - In southern California the United States Geological Service, for the first time in history, has issued a warning that a major magnitude quake is imminent in the short term.[9]

    - On the western coastlines of North America, the number of seismic events registering 1 and 2 on the Richter Scale have become so numerous that it is impossible to discern the end of one and the beginning of the next; it appears as one continuous rumble. The United States Geological Survey (USGS) has stopped issuing reports of seismic activity for level 1 and level 2 events.

- A marked increase in the number and diversity of miraculous healings occurring within individuals that the medical establishment had given up on such as cancer, Aids and EBV cases that the medical community had diagnosed as "incurable" are now in remission.[10] Many of these individuals have adopted changes in belief systems and lifestyle including a raw foods diet, meditation and prayer.

- New forms of bacterial infections that are resistant to every drug and antibiotic known at present. In 1992, 13,300 hospital patients died of infections from these bacteria.[11]

- Record numbers of new intentional communities forming outside the bounds of established community systems. These communities appear to be concentrated in Southern Colorado, Northern New Mexico, Oklahoma, Southern Texas and Northern California.

- Named the "Cosmic Event of the Century," the opportunity to witness, real-time, the collision of 21 cometary fragments into the planetary gas giant, Jupiter. Though uncertain as to the degree of the repercussions, Earth continues to respond to the impact through magnetic and seismic anomalies, as well as a shift of atmospheric water vapor on Earth.

- A dramatic increase in the number, geographical distribution and complexity of Crop Circles. Recent glyphs (1991–1994), appearing in cereal grains throughout the world, are irrefutable images based upon mathematical constants, sacred and ancient symbols of the indigenous populations, genetic information and Sacred Geometric patterns. As of late 1996, Crop Circles have been reported in every nation that grows cereal grain crops.

Obviously, life on Earth is not business as usual! As diverse as they may appear, all of these events are related. All are happening now. All are by-products of something much more significant than any one event alone. The common element underlying each of the phenomena is that of change. We see the change as shifting patterns of energy. What do these changes mean? Do they have reason and purpose? The answer is a resounding *yes* and you are a part of both the reason and the purpose.

Events such as these are becoming common, events that many said could never happen. Humankind is recognizing the need to reorganize itself into new patterns of expression. Traditional structures that are not in harmony with the emerging patterns of thought are falling away, making way for a more stable framework that is harmonious with the new thought. The world that is being experienced now will never be the same as the world of previous generations. It cannot. The change of many systems is appearing as massive breakdowns. We are witnessing systems stressed to the point of failure and the failures converging upon this point in time. Why are they occurring? Why now?

The events listed above are more than mere coincidences of events focused within a 30-year time frame. Please view these events through your own eyes, with your own heart and without the bias of mass media or consensus opinion. The world that you have always known, your world, is changing very rapidly. The structural guidelines that determine the way you think, perceive and act toward yourself and others are changing not only around you. They are changing within you!

THOSE WHO KNOW

The process that is occurring upon Earth in this moment, to the living information of each cell within your body, is so empowering by its very nature that the ancient texts providing knowledge of its existence were hidden in secluded libraries for safe keeping. This view of our yet to be lived history was relegated into obscurity within the esoteric teachings, secret orders and ancient mystery schools. The physical processes and their significance to us today have been the underlying thread of religious teachings, sacred orders and mystic sects throughout ancient as well as modern history. The Judeo–Christian traditions, Kabalistic teachings, the Orders of the Essene Masters, The Rosy Cross (Rosicrucians), The Golden Dawn, The Emerald Cross, The Amethystines and many more all have their roots in a single, fundamental body of information relating human experience to this time in history. Individuals have lost their lives in pursuit of understanding the profound nature of this change and this time in history. The events concerning today have been prophesied, predicted, and welcomed as well as ridiculed, scoffed at, termed "unbelievable," "impossible," and labeled as myth and folklore.

There may be those who never realize what has happened. In denial even now, they refuse to see the continuity of events that are unfolding before them daily, at a pace that warrants recognition. Having allowed themselves to become desensitized they listen, but do not hear the guidance that comes to them from within their own bodies. They look, but do not recognize, the patterns of change within their world. They await reassurances that life continues, business as usual. At the opposite end of the spectrum there are those who know, those who remember.

Taking the time to read these words is a clear indication that you have drawn this information into your life in an effort to understand, as well as remember. Regardless of your background, belief systems or prejudices, if the information within this text is meaningful to you at all, it will touch you deeply. As you read *Awakening to Zero Point: The Collective Initiation* you will experience feeling about what you are reading. Those feelings are your tools to becoming a different person. Through feeling you expand your awareness of the events unfolding within your world. Additionally, through feeling you expand your capacity of Compassion for yourself and others as we experience life together in the presence of one another. These days are the days that we have waited for, longed for. This time in history is the time that every experience, each relationship, within every lifetime has prepared us for. Without doubt, without hesitation, the events occurring now, within these days, are the reason that we have come to this world. We are living and witnessing a rare event by any standards, the birth of a New World. The events of our lives are the labor that is enabling this birth.

SACRED SIGNS

Within the pages of this text you will find keys to understanding the processes of creation, processes that you already know on some level. You had to have known them to masterfully project yourself through the portals of creation, through the dimensional tubes and into the womb of your mother. Occurring around you and throughout you, the processes are you! There is no difference, no separation between you and your world. Within you are all possibilities. You already know that you are a part of all that is, all that you may touch, feel, taste, hear, see, create or uncreate. You have been reminded that you are a part of all that has ever been. A being of infinite power, you exist as a focus of vast amounts of multispectral energy, with one portion of that energy focused as an individualized expression of human. As human you provide a unique sequence of experiences for yourself so that you may know yourself in all ways. You call these experiences your life.

Coming into this life you may have suspected that it would be difficult to transcend the illusions of gravity and magnetics. You may have known of the possibility that your experience would be challenging. You may have even suspected that it would become difficult to remember who you are, your purpose, your very nature. Perhaps most importantly, you may have suspected that you would grow to question your origins and connection with your own Creator. While you may have known, no one ever told you how great the possibilities. No one could because no one had ever experienced the journey that you have embarked upon.

Knowing these things, you came to experience and share your uniqueness with those that you love and hold most dear. Perhaps you knew that although you would be in the presence of others, your journey would be completed on your own and that someday, you would remember. Perhaps you have always known that there would be signs: sacred symbols, meaningful forms and patterns that would await you at some distant time when you would need them most. Among the symbols and signs there would be words to awaken your memory at the proper time. You trusted that you would recognize them. Possibly, within the pages of this text, you will find those words. Possibly, coded as resonant symbols of word and form, you will find triggers activating your ancient/future memory.

Perhaps these are the times. Perhaps, just perhaps, contained within these pages are symbols, the words and the images planted within your memory long ago to be accessed and awakened within you. It is my prayer that this material become a meaningful part of your day-to-day experience, bringing to you the peace that comes from the knowledge of purpose. I offer this information with great respect for you, individually and collectively, and your choice to experience *Awakening to Zero Point*.

CHAPTER ONE

The Shift of The Ages

A CHANGE-OF-STATE

No man can reveal to you aught but that which
already lies half asleep in the dawning of your knowledge.

For the vision of one man lends not its wings to another man.
And even as each of you stands alone in God's knowledge,
so must each one of you be alone in his knowledge of God
and in his understanding of the Earth.

ADAPTED FROM *THE PROPHET*, BY KAHLIL GIBRAN

There is a process of unprecedented change unfolding upon the Earth, now, within our lifetime. Without knowledge of the artificial boundaries of religion, science or ancient mystic traditions we see the shift as dramatic changes within the Earth, as well as within human perception and experience. This time is historically referred to as The Shift of the Ages or simply "The Shift." The time of The Shift marks our completion of a paradigm that has perpetuated the illusion of separation between ourselves and the creative forces of our world, and the birth of a new paradigm allowing the recognition of the oneness of all life.

The effects of The Shift of the Ages are reverberating throughout each and every aspect of creation, mirrored in all biological life that exists within the protective envelope of the Earth sphere. Every cell within each life form is restructuring its biochemical circuitry to generate, sustain and assimilate higher frequencies and more complex arrays of radiant information that we refer to as light.

Signaling the completion of a two hundred thousand year cycle of experience, The Shift heralds the beginning of the single most significant event to occur in conscious human memory. The rapid decrease in the intensity of planetary magnetic fields, accompanied by Earth's ability to sustain higher harmonics of her base pulse, mark the beginning of a new paradigm in conscious human awareness. As we witness the years of transition between "that which has always been and that which is yet to be," truly we are living our own birth to a higher order of expression. Our bodies are being asked to restructure their physical and morphogenetic fields of energy, information and light to accommodate the new codes of geometry activated by The Shift. Our restructuring and realignment results in profound effects; radical shifts of thought, feeling and emotion.

For some, this effect is experienced as the realization of those things that they would least choose to bring into their lives, their worst fears. By allowing the fears to come into conscious awareness, beliefs surrounding the fears may be examined, experienced and balanced (healed) gracefully. All that you have ever known, felt, touched, created or uncreated is undergoing a process that will forever change this world and the manner in which you perceive this world. Without exception, all that you have experienced throughout this and other lifetimes has been in support of this time, the time of The Shift.

The event of The Shift marks the closing of the current cycle of conscious evolution. Predicted, prophesied, celebrated and feared for thousands of years, entire religions, belief systems, sects, cults and orders have grown surrounding varied interpretations of this time in history. Referred to in many terms, the words themselves provide insight to the frame of reference of the observer-participant as well as the level and degree of understanding. A sampling, by no means complete, follows:

- The New Age
- The Second Coming
- Armageddon
- Polar Reversal
- Planetary Resurrection
- The Closing of a Cycle
- The Tribulation
- Fourth Dimension
- The Sixth World
- Planetary Pentecost
- Dimensional Translation
- The Age of Aquarius
- The Days of Judgment
- The Rapture

All of the events listed are related as different aspects of an identical process. All are by-products of something much more significant than any one event alone. There are books, seminars, workshops, lectures, study groups and retreats examining the nature of, understanding the purpose and preparing for some event described by one or a combination of, the terms identified above. For the purposes of this book, I prefer to use still an additional term, the Christing process, or simply the Christing. This is a nonreligious term describing the full expression of human potential. Without reference to any single belief system, the term *Christ* is used as the name of the highly evolved universal reference beings that have come to this world, the most recent nearly 2,000 years ago, to offer insights into the process that is culminating at this time.

History Points to Now

Ancient systems of time keeping indicate that today, this lifetime, is a very special time in the history of humanity as well as the planet. Almost universally, the calendars point to now as both the end and the beginning of a cycle of experience. Many of the Native American traditions, including those of the Lakota, Cherokee and Hopi indicate that this time in history is the time of their prophecies, the close of a grand cycle leading to the birth of a new world.[1]

The secret Hopi traditions, kept within tribal circles for centuries, in 1979 were released to the world by a Council of Hopi elders in northern Arizona. These traditions relate four previous worlds, the fifth of which we are in at present, and the coming of a sixth world of consciousness.[2] Each of the previous worlds ended at a point when all of humanity became "lost" and the separation between the heart and the mind became so great that only through a collapse of the lost world could a new world of order and harmony become reestablished. The separation was accompanied by catastrophic changes upon the Earth. For example, one cycle ended in fire, one in ice and one in torrential rainfall resulting in what is believed to have been the Biblical flood, providing a cleansing in preparation for the new world to emerge. It is the feeling of the Hopi Elders that the ancient Hopi prophecy has tremendous relevance to the events of today's world, and that all of humanity may benefit from the wisdom of the Hopi way of peace.

In a similar fashion, the Aztec traditions relate the four past "Suns" of Earth's history. Each sun depicts an epoch ending in a period of tremendous change. The First Sun, named the Nahui Ocelotl, was a time when our world was inhabited by giants living within the Earth. This period ended when the animal kingdom overcame the human kingdom. There are no indications of survivors of this epoch. The Second Sun, named Nahui Ehecatl, was noted as the time that new humans began to cultivate and crossbreed plants. The completion of this period was marked by great winds that swept the surface of the Earth clean. During the Third Sun, Nahui Quiauhuitl, Earth's populations constructed great temples and cities. Great openings within the Earth and a "rain of fire" marked the end of this cycle. Earth's Fourth Sun was an age of water. The inhabitants of Earth learned to navigate the great oceans. The Fourth Sun came to a close when tremendous floods covered the Earth making way for the Fifth Sun, which we are living today. The prophecies are unclear as to how the Fifth Sun ends. These traditions are indicated clearly on the face of the Aztec Calendar.

The Mayan Calendar is based upon Great Cycles 5,239 years in length, with each year consisting of 360 days. This ancient system of time keeping, by some estimates extending back 18,630 years, indicates that the present day is very significant in terms of Earth, as well as man. As Jose Arguelles points out in his work, *The Mayan Factor*, 1992 marks the beginning of the last *subcycle* within this great cycle.[3] This portion of the Calendar is predicted to last for 20 years, finding completion in the year 2012 A.D. At this point the Great Cycle of the Mayan Calendar comes to a close, marking the beginning of a new cycle; a cycle for which there is no calendar at present. The Mayan system of time keeping, dating to 18,000 years ago, *predicted* that this period in history would be the lifetime of change.

Plato, the Greek philosopher, recounts that when his ancestor, Solon, met with high priests in ancient Egypt they told of records indicating previous cycles of destruction and birth. Each cycle was accompanied by "inclinations and disturbances" of heavenly bodies, including the Earth.This information is supported in the lost Biblical *Book of Enoch* (not to be confused with the present-day Keys of Enoch). Chapter LXIV of the *Book of Enoch* speaks of a time when "Earth became inclined upon its axis" just prior to the great flood. Enoch related to his son, Methuselah, visions of ancient events followed by his prophetic vision of events yet to come.

> *...and everything done on Earth shall be subverted and disappear in its season.... In those days the fruits of the Earth shall not flourish in their season,... heaven shall stand still. The moon shall change its laws, and not be seen at its proper period.*
>
> ADAPTED FROM THE *BOOK OF ENOCH*[4]

The former heaven shall depart and pass away, a new heaven shall appear.

ADAPTED FROM THE *BOOK OF ENOCH*[5]

Several passages of our modern Bible correlate highly, though less completely, regarding the time of The Shift.

There shall be famines and earthquakes in divers places...great tribulation, such was not since the beginning of the world to this time, no, nor shall ever be...Immediately after the tribulation of those days, the sun shall be darkened, and the moon shall not give her light, and the stars shall fall from heaven...

ADAPTED FROM *THE HOLY BIBLE*,[6] MATTHEW 24:7

I saw a new heaven and a new Earth, for the first heaven and the first Earth had passed away.

ADAPTED FROM *THE HOLY BIBLE*,[7] REVELATION 19:11

The Emerald Tablets of Thoth, translated through several languages, date to approximately 39,000 B.C. The tablets are not channeled material, rather the direct words of Thoth recorded upon 12 crystalline tablets in an effort to preserve the knowledge, heritage and mysteries of the human experience within this cycle. The tablets were reportedly last seen in 1925 and have become the basis for many of the teachings of Hermetic Science, as well as the doctrine for the Order of the Free Masons. Repeatedly the tablets emphasize that the human form is undergoing a "Shift," and in doing so, is learning a new expression of the relationship between "Dark" and "Light":

Man is in process of changing to forms of light that are not of this world; Grows he in time to the formless, a plane on the cycle above. Know ye, ye must become formless before ye are one with the light.

ADAPTED FROM *THE EMERALD TABLETS OF THOTH*[8]

When man again shall conquer the ocean, and fly in the air on wings like the birds, when he has learned to harness the lightening, then shall the time of warfare begin. Great shall the battle be twixt the forces, great the warfare of darkness and light. Nation shall rise against nation, using the dark forces to shatter the Earth.
...Then shall come forth the Sons of the Morning, and give their edict to the children of men, saying: O men, cease from thy striving against thy brother, only thus can ye come to the light...Then shall the age of light be unfolded, with all men seeking the light of the goal.

ADAPTED FROM *THE EMERALD TABLETS OF THOTH*[9]

THE SHIFT: OUR RETURN

The nature of The Shift has been questioned, pondered, postulated, hypothesized and worshipped for thousands of years. Religions have developed surrounding what appear to be well-intentioned, though distorted, understandings of this elusive yet fundamental force of creation. The consequences of this Shift transcend the boundaries of religion, science and mysticism. All are languages developed throughout history to understand creation, creative processes, the origin and ultimate fate of human life. Each represents a single portion of a much larger, more all-encompassing truth.

The process of The Shift itself may be viewed as analogous to that of a very familiar shift seen on a daily basis: the *change-of-state* or phase transitions of water. Water may be seen in any one, or a combination of, three forms or states as they are referred to: solid (ice), liquid (flowing water), or gas (vapor or steam). Chemically each of these forms of water is the same, H_2O. Structurally, the geometry of the molecular packing is different, allowing the compound of water to express differently under varying conditions. Illustrated through the use of a particular graph known as the Phase Diagram (Fig. 1), the ability of water to express itself within each one, or a combination of, three states is shown as a function of temperature and pressure.

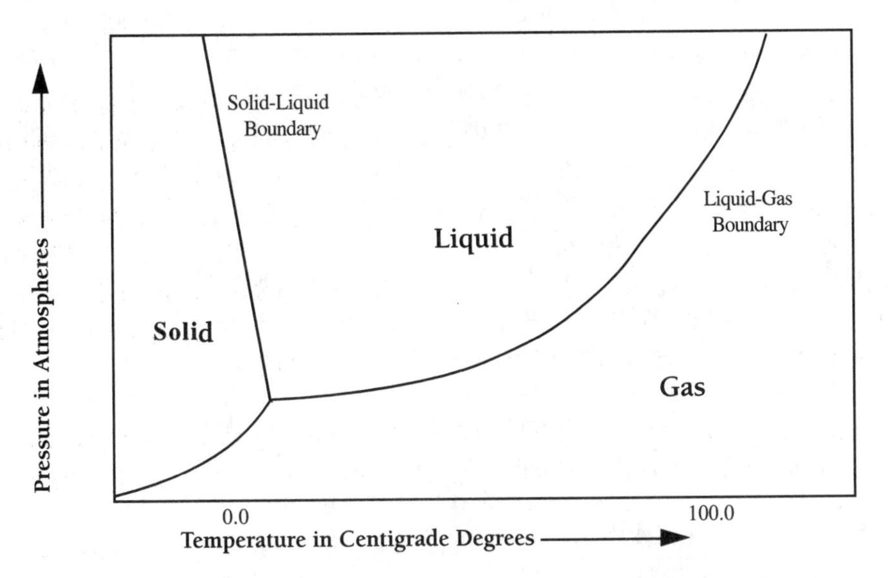

Figure 1 *Phase diagram showing water as three states: solid, liquid and vapor.*

This particular phase diagram is concerned with events happening in two directions, represented by two axes, one of X and one of Y. In this diagram the X axis, shown as the horizontal line, represents the parameter of temperature and increases from left to right. The second axis, shown as the vertical line Y, represents atmospheric pressure and increases from bottom to top. In general, the illustration indicates the following:

- At low temperatures within a wide range of atmospheric conditions, water expresses itself as a solid, ice. Chemically it is still water, H_2O. Structurally it has become more dense; the molecules are moving very slowly.
- As temperature increases and atmospheric pressure increases, water begins to become less dense and begins to express itself as gas or vapor (steam). Chemically it is still water. Structurally the molecules are moving more rapidly.

- As temperatures decrease and atmospheric pressures increase water is able to express itself as flowing liquid, water. Again, chemically the same. Structurally, molecules are more mobile than in ice, however, less mobile than in vapor.
- There is one very special point in this diagram, represented as the intersection of all phase boundary lines in the center of the graph. The point is named the "triple point" and is the one point representing the one set of criteria within which water may exist simultaneously as any of the three phases: solid, gas or flowing liquid.

Chemically the water remains the same, H_2O, while structurally it expresses itself differently as a function of its environment, in this instance temperature and pressure.

Further examples of matter demonstrating a change-of-state may be found within the mineral kingdom. Many minerals exhibit varying external expressions, while retaining the chemical properties that allow them their identity by definition. For example, the mineral fluorite is commonly found in large clusters of perfect cubes bordered upon its flat faces by other cubes. In the same deposit, fluorite may also be found expressing as the geometric form of the octahedron, an eight-sided crystal reminiscent of two four-sided pyramids mirroring one another with the widest portion, the base, the plane of reflection. Chemically, both forms are identical, CaF_2. Structurally, they express differently as a function of the processes that created them.

Iron pyrite may be found in the field as discrete hexahedrons (perfect cubes) loose in the deposits or bound to other cubes. Again, in the same deposits, the pyrite may also express as the geometric form known as the dodecahedron, an approximation of the sphere with 12 faces. Chemically the same, FeS_2, structurally different. Conceptually, the phase diagram of water and the mineral examples are very accurate metaphors for the process of the planetary shift being experienced by us and our planet.

Chemically, Earth will remain the same through The Shift of the Ages. Structurally, it is the morphogenetic expression that is reflecting The Shift. The "environmental"* parameters of Earth are changing. Rather than temperature and pressure in our water example, our experience is governed, in part, by planetary magnetics and frequency. As these parameters shift, in much the same manner as the temperature and pressure of the phase diagram for water, creation accommodates these changes by expressing itself in a different manner. Chemically, matter remains the same. Structurally there is change (Fig. 2). Conceptually the human form may be considered as one composite compound, such as water, within a dynamic and multivariable, evolutionary environment. This evolutionary environment has been predominantly governed by planetary fields of magnetism and frequency. Catastrophic events throughout history have had a profound impact on human evolution through the interruption of these fields.

* For the purposes of this text, the term "environment" is a reference to fields of energy affecting the cellular complexes of life on earth. These fields are primarily related to magnetics and base resonant frequency, or the fundamental vibration of earth, and not environmental conditions of weather patterns, temperature fluctuations and ozone or greenhouse effects, though these are all related to the shift patterns of discussion.

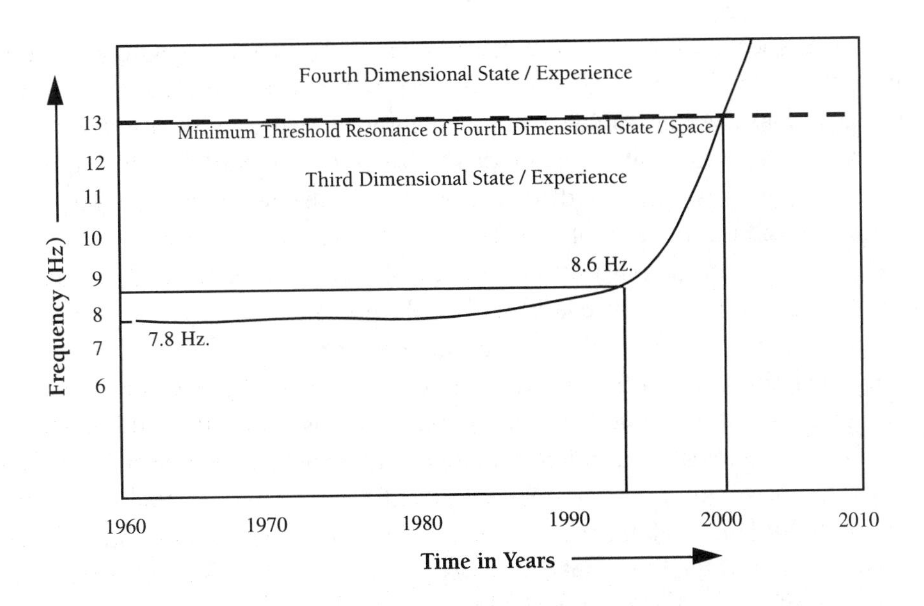

Figure 2 *Idealized phase diagram indicating relationship between Earth's frequency, and dimensional expression over time. The dates beyond 1994 are approximations only, within an ideal state, and are not to be viewed as prediction.*

For approximately the last two hundred thousand years we assume that Earth has functioned within one specific zone, or range, of frequency. We "assume" because we have only measured these fields directly within the last 100 years. All matter, all that human consciousness has known, felt, touched, created or uncreated has occurred within the context of matter expressing within this range of frequency. Living within the radiant fields of Earth, all life has been in primary resonance with this key zone (Fig. 3).

There has always been another range of frequency, or band of information, that has always lived within the range of our fields. This body of information has appeared

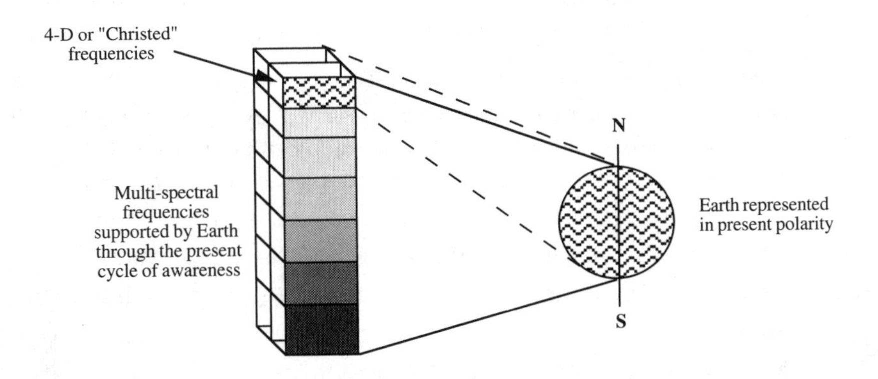

Figure 3 *Until the present, Earth has supported multiple zones of information as ranges of frequency. Human experience within the last 200,000 years of this conscious cycle has been within the context of matter expressing one, or some combination, of these frequencies.*

as a zone of higher frequency that has remained available, though possibly appearing less accessible, to each individual. It is into this new range of highly evolved information that both Earth and humankind are moving into exclusive resonance. It is to these frequencies that each cell within our bodies is attempting to "map" itself. Our migration into complete resonance with this new body of information, Resurrection followed by Ascension, is the goal of The Shift.

Our tuning to this highly evolved band of information is not automatic. Any relationship to this body of knowledge is one which must be *achieved*, using the tools of choice and free will applied to the processes of life. It is into the zone of experience, offered through this new range of frequency, that the phase transition will carry Earth. A function of conscious and intentional evolution, Earth will no longer support inharmonic patterns of fear, hate, the polarity of judgment, or the belief systems of an obsolete paradigm. Earth will support one, highly evolved body of information, that may be thought of as the Christed frequencies (Fig. 4). It is to this zone of information that the one known as "Jesus," the Universal Christ, was in resonance. This was the gift of Christ, to anchor the information of this consciousness, and all of its possibilities, firmly into the conscious matrix of humankind. He accomplished his goal through the expression of his life in our presence. We saw him among us as he lived his truth before our eyes, assuring that we would notice and remember. The process of his life became the living bridge, making the higher range of information accessible to all of human kind.

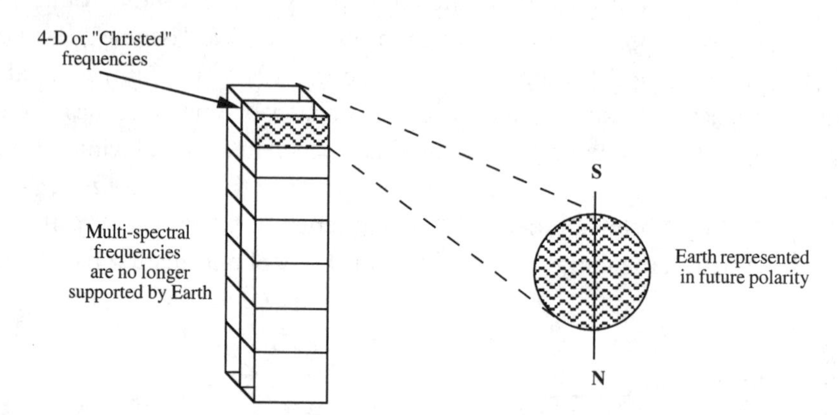

Figure 4 *Earth is completing an evolutionary migration away from the multispectral band of lower frequency information, to a single band of highly evolved, high-frequency information; the Christed frequencies. All life harmonically "tuned" to Earth will be in exclusive resonance with the new frequencies following The Shift.*

A Western physicist might describe this initial shifting of resonance as a dimensional shift. In Biblical terms, the act of consciously vibrating from one state-space into another is referred to as Resurrection. We may define Resurrection as our allowing the mind, body, spirit complex to harmoniously shift to a higher expression of itself rather than give in to the urge to separate as death. The living example of the universal Christ demonstrated, that through the conscious use of choice and free will, man in his totality is greater than the fragments of his fear or perceived limitations.

The Shift of the Ages is the term applied to the process of Earth accelerating through a course of evolutionary change, with the human species, linked by choice, to the electromagnetic fields of Earth following suit through a process of cellular change. This process will be discussed in detail throughout subsequent sections. The human aspect of The Shift may be consciously facilitated, even accelerated, through the use of choice, free will and emotion associated with the ancient wisdom of the human mind, body, spirit relationship. This is the purpose of The Shift, the ultimate balance and healing of Earth and all life forms that are capable of sustaining the energy of that healing. This is the shift to a new way of expressing the human form, through the lens of higher frequency; a Christed energy. This is *Awakening to Zero Point: The Collective Initiation.*

THE NATURE OF STILLNESS: ZERO POINT

In traditional physics there is a general assumption that "things may happen" only in a space where there is the absence of a vacuum. It is within this space that the forces of temperature and pressures drive the systems of creation, producing events that may be observed and measured. Scientists typically use this principle embodied as an instrument to measure temperature in the laboratory. A glass thermometer indicates temperature through the rise and fall of a column of mercury within the vacuum of the sealed tube. As temperature decreases, the pressure of the gas within the tube decreases correspondingly. In theory, there is a point at which the pressure would drop to zero, with a corresponding drop in temperature yielding zero degrees on a scale known as the Kelvin Scale. An absolute scale of temperature, the Kelvin Scale zero point is found at -273.15 degrees C. It is at this point, in principle, that all molecules are at rest and become "still" as gases exert no pressure and occupy no volume (Fig. 5). Another term for this point would be that of absolute zero.

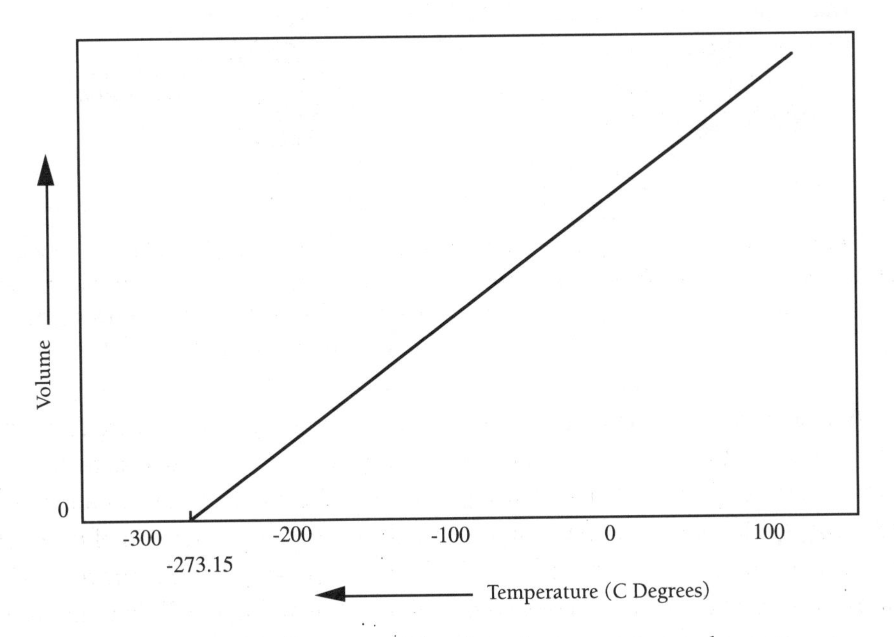

Figure 5 *Graphic illustration of Zero Point for temperature and pressure.*

Newton's third law of thermodynamics states that at absolute zero, all molecules are perfectly aligned and motionless, with the degree of disorder (entropy) at zero. Traditional physics states that this is a theoretical point only, not possible to attain experimentally. *Quantum physics, however, predicts and allows for continued motion through absolute zero.* Following the guidelines of Newton's third law traditional science has accepted the theory that it is not possible to achieve zero degrees Kelvin. A specialized study of physics, quantum physics, both allows for and predicts fluctuations within a vacuum down to and including zero degrees Kelvin. The point at which temperatures reach absolute zero, with a corresponding decrease of pressure, is referred to as Zero Point; the amount of vibrational energy associated with matter at zero degrees Kelvin.

It is in the space of zero point that creation becomes very quiet and still
to the observer, though energy is still in motion
within the vacuum and experienced by the participant.

Earth is experiencing early stages of the events that will provide an experience of Zero Point, allowing the breakdown of thought-constructs that are inharmonic with our blueprint of expression. In place of temperature and pressure, the parameters of magnetics and frequency are providing the conditions necessary to achieve Zero Point. Each individual now living upon the Earth is an integral part of The Shift process, playing the vital role of midwife in the birthing of a new era of human perception and awareness.

The intent of presenting *Awakening to Zero Point* at this time in this particular format is to empower the reader through knowledge of the mechanisms of The Shift process; the inner workings of the event on a conceptual level. We have been inundated for centuries with prophecy, prediction and warnings of catastrophic change within this period of history. We have been asked to accept and believe the information as it has been presented *to* us, as we have been *told*. The following sections will provide the mechanism to *know* from within, of the change that is occurring, and present it in terms relevant to daily life. In this manner it becomes apparent why the changes are occurring and the interrelated nature of various aspects of the phenomenon. The experience of Zero Point is the goal of ancient meditative practices and is closely related to the Biblical term of Resurrection, as well as, The Shift of the Ages.

The Shift is far from a hypothetical event to be viewed as something to occur during some distant geological epoch far into the future. It is beyond a process reserved for mystical, esoteric theoreticians living recluse lives in remote locations of the globe waiting for the end of the world as we know it. The Shift is a sequence of knowable, measurable processes and events that are already under way.

The Shift of the Ages has already begun!

MAGNETICS: THE FIRST KEY OF RESURRECTION

The Shift may be addressed from a variety of perspectives, each a valid language in its own right. Esoteric discussions will center around "moving toward the light" and the coming of the "New Age." An equally valid language that may be used is that of Earth's changing physical dynamics. From the perspective of earth science, the changing paradigm is accomplished through a realignment of two digitally measurable, fundamental parameters: those of planetary frequency and planetary magnetics. These parameters alone have a far-reaching and profound impact upon human consciousness, human thought and perception specifically, and the behavior of matter in general. Both of these parameters are changing at present, having fluctuated dramatically in both the historical, as well as, the geological past. Each has a dramatic yet subtle effect upon the cellular body, human consciousness, and the manner in which that consciousness expresses itself.

In the college textbook, *Physical Geology* by Leet and Judson it is stated:
"The cause of the Earth's magnetism has remained one of the most vexing problems of Earth study. A completely satisfactory answer to the question is still forthcoming." [10]

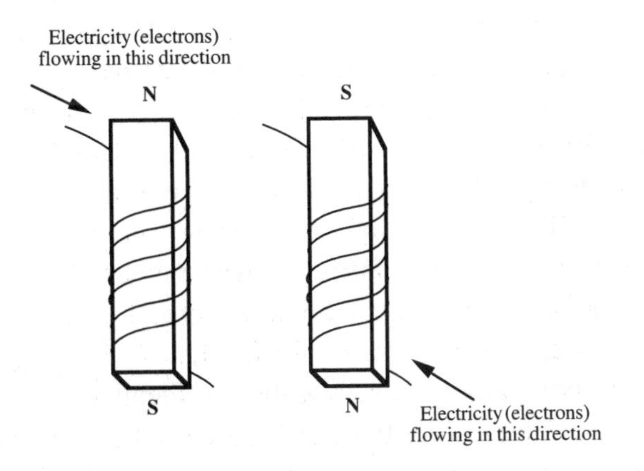

Figure 6 *An electrical charge passed in a circular fashion around an iron bar creates a magnetic field with polar expressions of both North and South. A reversal of the direction of the electrical charge creates another magnetic field, with the sense of the field reversed, though the orientation of the bars has not changed.*

The relationship between Earth, the planetary magnetic fields, and cellular function of the body is a key component to the understanding of the conscious evolution and the process of The Shift. A conceptual knowledge of the magnetic fields of Earth may be gained through the use of a simple demonstration. Consider a simple iron bar of any dimensions. The iron appears as dense with no magnetic properties. If an arbitrary length of conductive wire is wrapped around the iron bar in one direction, with any number of windings, and an electrical charge is passed through the wire in an arbitrary direction, an interesting phenomenon occurs. The previously nonmagnetic iron bar becomes magnetized, developing a magnetic field with polar expression of North and South (Fig. 6).

The next portion of this demonstration produces a very significant, and possibly unexpected, effect related to the shift process of Earth. As the flow of the electrical charge around the iron bar is reversed, with the bar remaining in its original position, the first magnetic field is lost and a second magnetic field is generated in its place. This second field, however, has significantly changed the manner in which it is expressing the magnetic effect; *it is now reversed.* Without changing the physical orientation of the bar, what was once the Northernmost pole on the bar has become the Southernmost and the Southernmost pole is now the Northernmost. Simply by altering the *direction* of the flow of electrons relative to the iron bar, the *sense* of the magnetic fields has reversed while the bar itself has remained in its original position! If this field is generated upon a flat surface of iron filings, the individual particles of iron will align themselves along the arc line of the magnetic field, so that the arc becomes visible. The lines of force on the flat surface will look like the lines of force that surround Earth (Fig. 8).

The key to the field of magnetics may be found in the motion of the electrons themselves. Electrons moving in a circular motion around a relatively fixed body of iron produce the phenomenon of magnetics. The electrons of electricity are being guided along the pathway of the conductive coil, in a circular motion around the iron bar, generating a new field of force at a 90 degree angle to the direction of the electron flow. This demonstration is a very good analogy for the physical dynamics of The Shift processes that are occurring upon Earth at present.

A generalized cross section of Earth will reveal that the planet is not uniform throughout (Fig. 7). Rather, it is composed of layers of material creating zones varying in temperature and density. Each of these parameters is a function of depth below surface and tremendous pressures associated with these depths. The outermost layer is referred to as the crust and provides the visible surface of the continents and oceans. The crust is a relatively thin layer averaging three miles thick beneath the oceans and 25 miles in thickness as measured through the continents.

Below the Earth's crust there is a second layer, averaging approximately 1,800 miles in thickness. This is the Earth's mantle. The material of the mantle is much denser than that of the crust and exists at such high temperatures and pressures that it is essentially a thick molten liquid. It is this plastic material from the mantle that extrudes itself onto the surface through openings in the crust such as volcanoes and lava flows.

Under the mantle material lies a thinner, yet more dense zone of material, the core. Scientists divide the core into two zones, the outer and inner core, estimated at 1,366 and 782 miles in thickness, respectively. The inner core is believed to be a plastic-like sphere, with the outer core a viscous liquid of molten material, warmer than the core but cooler than the next mantle layer.

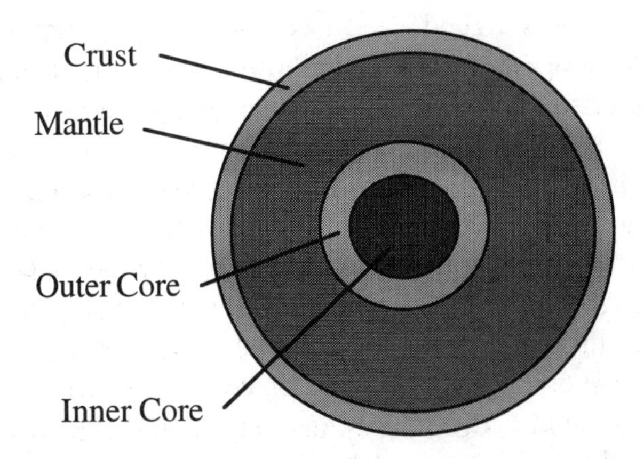

Figure 7 *Schematic view of Earth in cross section showing the relative thicknesses of the primary zones of crust, mantle and the molten core of iron and nickel.*

Within the iron and nickel core of Earth, our iron bar analogy resumes. In the demonstration (Fig. 6) the movement of the electrons around a stationary core of iron produce the effect of the magnetic field. Earth's predominant field of magnetics has a simple dipolar shape, as if the planet had a huge bar magnet as its core. It is the rotation of Earth around the molten inner core, that generates an excess of electrons (-e) within the crustal layers. Following the laws of classical physics, a proportional field of magnetics is generated at right angles to the flow of electrons, yielding the shape of the donut-like magnetic fields.

These fields, measured in units known as gauss,* are a function of the composite rotation of Earth around these iron cores in general, and specifically the motion of the outer core relative to the inner core.

The more rapid the rotation, the greater the intensity of the magnetic field around the planet's iron and nickel core.

The movement of rotation is analogous to the electrical charge moving in a circular motion through a conductor wrapped around the iron bar (Fig. 8). For this discussion, the mode of transport for the electron stream is not significant, be it the charge of planetary rotation or conductive copper winding. In each example, electrons are moving around a relatively stationary source of iron producing a field effect that is termed magnetic. The implication is that the intensity of the fields of magnetism is a function of the rate of rotation. The more rapidly Earth rotates, the denser the fields of magnetism become. Conversely, the slower the planetary rotation, the less dense the fields of magnetism. This is precisely what is happening at present, as well as what has happened numerous times throughout planetary history. Though alluded to in the ancient texts, evidence supporting the rotation/magnetic relationship is just beginning to surface in recent years. In a paper published in Sweden in 1988, Nils-Axel Morner reports that

* After Karl Gauss, the unit used to measure magnetic flux density

"because the inner core is a good electric conductor and carries a large rotational energy, it is likely to have a strong interaction with the main geomagnetic field."[11]

In addition to the field of global magnetics, Earth's rotation within an envelope of a multilayered atmosphere produce an electrical charge that may be measured as "static potential"; that is the electrical charge builds to a certain value before it is discharged. Tesla discovered, and modern science now recognizes, that the Earth/atmosphere system essentially functions as a large spherical capacitor, with the surface (ground) negatively charged through an excess of electrons. Layers of the upper atmosphere exhibit a positive charge, creating an *electrical potential* with an average value of 130 volts per meter over the surface of the Earth. The phenomenon of lightening is a dramatic and beautiful demonstration of Earth's attempt to reach equilibrium, balancing the charge between the atmosphere or ground. Global discharges of lightening involve approximately 2,500 strikes over a 100 square kilometer area per year, as part of an ongoing attempt by Earth and the atmosphere to reach a perfect electrical balance.

You are bathed in this electrical potential throughout your life; seemingly without effect. It is this "trickle charge" of static electricity that is partially responsible for holding the magnetic alignment between your body and the patterns of your experience. Recent studies have validated the brain/Earth relationship through magnetics.

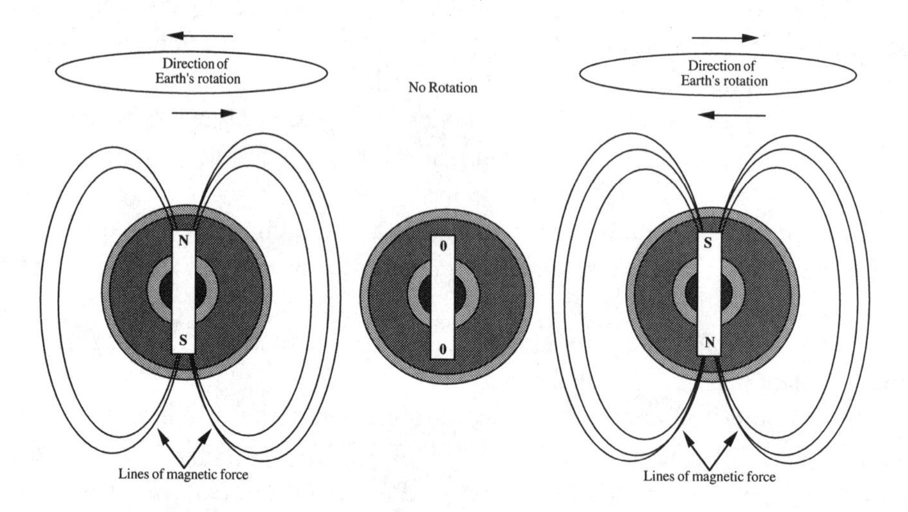

Figure 8 *Schematic illustration of the relationship between the rotation of Earth, the orientation of North and South poles, and the fields of magnetism that result from the rotation. In the absence of rotation, magnetic values decline to "0" and Earth experiences a period of magnetic "null" or Zero Point. As the rotation begins in the opposite direction, new magnetic fields develop in a polarity that is opposite to that before the null period.*

An international team studying the phenomenon of "magnetoreception," the ability of the body to detect magnetic fluctuations, announced that the human brain contains "millions of tiny magnetic particles" (*Science*, Vol. 260:1590, June 11, 1993). The ancient texts tell us that the body seeks a harmonic balance with the Earth. This balance is the goal of our life experience, and may be consciously regulated through nonpolarized thoughts of forgiveness and feelings of Compassion. The magnetic particles within our brain serve as a physical link; the static potential serves as the "trickle charge" holding the information component of the particles in place.

The geologic record indicates that the magnetic fields of Earth have shifted previously, at least 14 times in the last 4.5 million years,[12] as determined by magnetic measurements taken from extrusive rock cores (earth material that was once molten and has been ejected to the surface, cooled and preserved the alignment of minerals sensitive to magnetism) (Fig. 9). Additional evidence of a 180 degree polar shift relatively recently may be seen again in the work of Morner in a paper entitled *Earth's Rotation and Magnetism.*

> *Radiocarbon dates of carbonate concretions within the varves gave ages that suggested that the varves (sedimentary deposits) were laid down between about 13,600 and 12,800 BP (before present) with the transpolar shift occurring at about 13,200 BP. This means that the same transpolar shift is now also recorded from the Southern Hemisphere. This can hardly mean anything else than a dipolar nature of the shift.*[13] (Parentheses are the author's.)

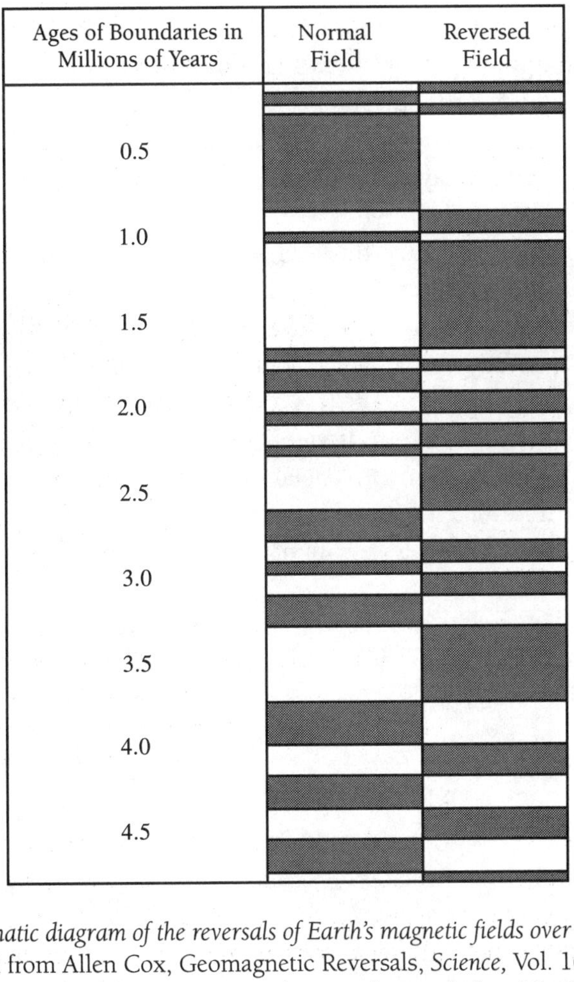

Ages of Boundaries in Millions of Years
Normal Field
Reversed Field
0.5
1.0
1.5
2.0
2.5
3.0
3.5
4.0
4.5

Figure 9 *Schematic diagram of the reversals of Earth's magnetic fields over the last 4.5 million years.* (Adapted from Allen Cox, Geomagnetic Reversals, *Science*, Vol. 168, Fig. 4, 1969.)
The geologic record illustrates 171 such reversals over the last 76 million years.

If the strength of Earth's magnetic fields is, in fact, a function of the core/mantle relationship, a reversal of the poles would seem to indicate that the motion of these bodies has slowed and reversed coinciding with the shift. If such an event did occur within the collective memory of humanity, would it not have been recorded? Possibly.

Apparently, there are records of at least one event, in the collective memory, of not one but two separate civilizations, when the rotation of the Earth exhibited a very unusual behavior. In his book, *The Lost Realms,* Zechariah Sitchin recounts narratives from these societies, one from the Peruvian Andes, another from Biblical texts, of anomalous activity regarding Earth's rotation. During the time of Titu Yupanqui Pachacuti II, approximately 1394 B.C., there was a time of anomalous night when,

"...there was no dawn for twenty hours."[14]

This event was not describing an eclipse, as none was recorded or predicted for that time by either Chinese or Peruvian astronomers. If the event were attributed to an eclipse, none has been known to last for such a long period of time. Something happened that was interpreted as immersing one portion of Earth in night for 20 hours, nearly twice as long as should be possible.

Sitchin theorizes that if such an event did occur, somewhere on the opposite side of the world an opposite event should have been recorded. The King James version of the Biblical texts records such an event in the Book of Joshua 10:13:

"And the Sun stood still, and the moon stayed, until the nation took vengeance on their enemies...The sun stayed in the midst of heaven, and did not hasten to go down for about a whole day."[15]

According to Biblical scholars, this event took place sometime shortly after 1393 B.C. Is this conclusive proof that the Earth routinely slows its rotation to a standstill? Certainly not. These accounts do indicate, however, that there are times within the memory of human consciousness (rather than geologic time) when the rotation of Earth has performed unusually. In a later portion of this section, you will see why the majority of records from a particular cycle do not survive The Shift. Here is a clue: The only materials that may survive the disruption of planetary magnetics and base resonant frequency are those that are Earth resonant. Regardless of what values Earth parameters shift into, the materials are always tuned to those parameters.

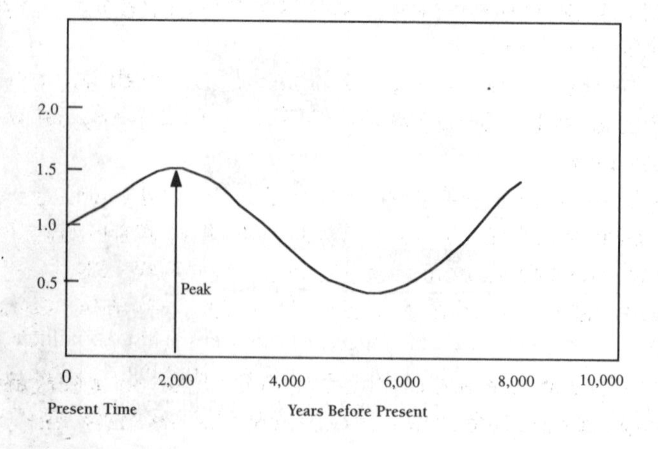

Figure 10
Schematic diagram of Earth showing magnetic fields in decline for the last 2,000 years. (Adapted from V. Bucha, *Archaeometry,* Vol. 10, p. 20, 1967.)

At present, the indicators preceding such a shift may not be readily recognized although they are acknowledged. In the June 1993 edition of *Science News,* an article concerning magnetic reversals states:

"The task of finding an accurate reversal record seems to be all the more difficult because the magnetic field weakens considerably when it switches direction."[16]

The intensity of Earth's magnetic field is dropping rapidly at the present time. Geologic records indicate that Earth is declining from a magnetic peak 2,000 years ago and the values have steadily dropped from that time to the present. The data indicates that the intensity of our planetary magnetic fields is approximately 38% lower than it was 2,000 years ago.

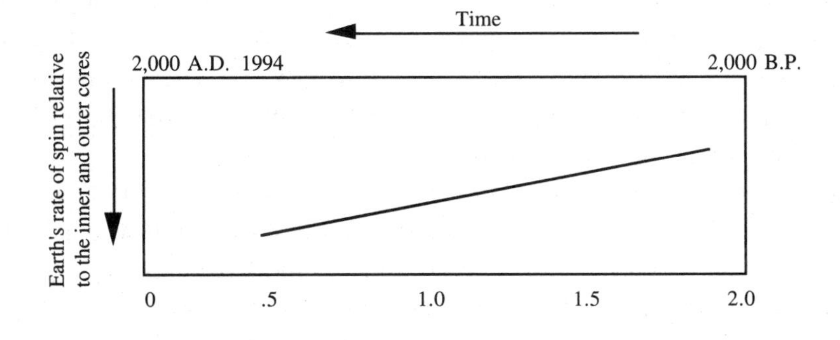

Table 1 *Earth's field of magnetics over time as the ratio of past intensity/present intensity. Graphic illustration of Earth's declining rate of spin (composite as well as inner/outer core coupling) and the decrease of magnetic intensity over time, as they approach "Zero Point." Values beyond 1994 are approximations.* (Adapted from V. Bucha, *Archaeometry,* Vol. 10, p. 20, 1967.)

Measurements over the last 130 years indicate a decline in magnetics from 8.5×10^{25} gauss units, to 8.0×10^{25} gauss units, or an average rate of about 6% per 100 years.[17] As magnetics are a function of planetary rotation, a lessening in the intensity of magnetics would seem to indicate a lessening in the rate of Earth's rotation. This is, in fact, precisely what is happening, within the inner and outer cores of Earth, as well as the overall rotation of the planet. In a 1985 report entitled "Solid Earth Magnetism," Tsuneji Rikitake and Yoshimori Honkura of Japan report,

"In association with a minimum in the westward drift velocity of the eccentric dipole, which appeared around 1910, *retardation in the rate of Earth's rotation was observed.*"[18] (Italics are the author's.)

Twice in 1992 and at least once in 1993 the National Bureau of Standards in Boulder, Colorado reset the cesium atomic clocks to reflect "lost time" in the day. Our days become longer than our clocks can account for. "If we didn't do this, we'd eventually get out of synch with sunlight," according to Dennis McCarthy, an astronomer at the U.S. Naval Observatory.

The effects of global magnetics are not confined to individuals on a personal level. Variable planetary magnetics provide zones of experience where mass units of consciousness are drawn to feel or work out some form of common experience. When an individual or group consciousness feels that an area no longer feels appropriate, or resonates with them, they are describing the relationship of their body's sensors to those zones of magnetic density. The understanding of the nature of these fields is perhaps a vital key to understanding mass migrations of large populations, human and animal alike, as well as the unexplained settling of ancient cultures in what may appear to be very unlikely locations for commerce or spiritual pursuits. Chaco Canyon in Northwestern New Mexico provides such an example. An interesting correlation may be drawn through the overlay of a global map indicating vertical field intensity of magnetics onto a political map of Earth at present. You will recall that lower values of magnetics provide the opportunity for change. How the change is expressed remains the choice of those within the experience. Conversely, readings of higher magnetics indicate zones where change is not as quick to occur. In some areas this may appear as stagnation. Please review Figure 11. Note the geographic locations and political zones that lay within one or two contours on either side of low magnetic zones. One of the most striking is that of the Middle East. A zero contour runs directly along the northeastern border of Africa, along the Sinai Peninsula through the Gulf of Suez and into the Red Sea; certainly an area that has struggled with tremendous change for centuries. Much of the change has been expressed as conflict leading to war and political turmoil. Even lower contours continue to the west into Egypt and the Sudan. The west coast of North America also displays magnetic contours of low values and is known for change of a different kind: new thought and innovation in the fields of technology, fashion and politics. This low contour continues South and approaches Central America, the location of rapid and often radical change brought about through political and military turmoil. Again, these zones simply offer the opportunity for change. How the opportunity is expressed becomes the choice of those experiencing the change.

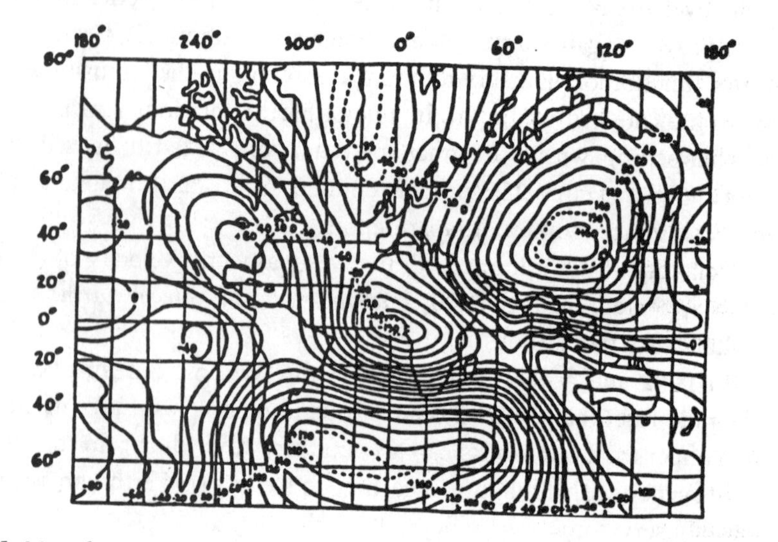

Figure 11 *Map of vertical magnetic field intensity in 1945 with contour intervals at .02 gauss.* (From Bullard, Freedman Gellman, and Nixon, *Phil. Trans. Royal Society of London, A, 243:67–92 [1950].)*

On the opposite end of the spectrum, there are zones of extremely high magnetics, yielding areas where change may not be as rapid or accepted. In Central Russia, for example, contour values read in excess of 150 gauss units for that portion of the country. The southeastern portion of the United Sates also exhibits relatively high values of magnetic intensity. Historically, these areas have been slower to respond to change, adapt new ideas and provide innovation.

The values of these contours are not permanently fixed. They are in transition. The "drift" of Earth's magnetic fields has been determined to be in a westward direction and a direct function of the rotation of the planet.[19] Furthermore, for the fields of magnetics to make one complete rotation around the surface of the Earth requires a time span, very familiar to you by now, of approximately 2,000 years. The symmetry of this process is magnificent. Sometime within the 2,000 year time frame, each continental land mass of Earth will have the opportunity to experience some aspect of Earth's magnetics. At the close of the 2,000 year subcycle each continent will have experienced a full compliment of the magnetic spectrum of intensities. This would indicate that now, less than ten years from the close of this cycle and the 2,000 year anniversary of the birth of the Universal Christ, the alignment of magnetic contours are nearly identical to those that were present at the birth of Jesus of Nazareth. Earth is poised to experience the same magnetic alignment now, on the threshold of a new paradigm of experience, with the identical alignment that ushered in the opportunity of a similar paradigm 2,000 years ago!

What, then, is the significance between fluctuating magnetic fields of Earth and the Awakening process of the evolution of human consciousness? To understand this it becomes necessary to develop a working knowledge of the relationship between human consciousness and the magnetic fields of Earth.

MAGNETIC TENSION: THE GLUE OF CONSCIOUSNESS

The energy referred to as consciousness is electromagnetic energy/information/light that is bound up within some aspect of the magnetic fields of our planet. Conscious essence may be considered as hierarchical grid upon hierarchical grid of this energy, forming continuous matrices of subtle frequency and geometry. It is within these matrices that magnetic influences provide a "tension" or stress field binding the essence of human consciousness as a framework of divine intelligence. These fields of information are bound to the Earth sphere through a stabilizing glue of planetary magnetics. Please understand that it is the *awareness* of humankind, and not the life essence itself, that is used to interpret the three dimensional world, the self and ultimately the Creator. In its simplicity, life essence has no need to understand. It is this awareness that is locked up within the fields of magnetics surrounding the planet. Through the structure provided as fields of magnetics the net of our awareness matrix is stabilized and secured in place.

You may view yourself as many things on many levels, and may be categorized and defined through the unique vocabulary used to describe yourself on each of those levels. Biologically you are bones, flesh, organs, cells, fluids, etc. Geometrically you are crystalline in nature. Each biological component of your body may be reduced to one, or some combination of crystalline substances. Bone, for example, reduces to the crystalline form of calcium. Blood reduces to the crystalline form of iron and trace minerals. If you were to be scanned electromagnetically, you would appear as a composite wave form, expressing yourself as a series of geometric patterns representing many individual wave forms from each unique biological aspect of your body.

Energetically, we are electrical in nature. Each cell within each component of our body generates a charge of approximately 1.17 volts at a specific frequency for that organ. This unique vibration is termed a *signature* frequency. Each cell is in constant motion, the rhythmic oscillation of a subtle beat, generating its signature frequency. We are more than simply electrical beings, however. We are both electrical and magnetic in our Earthly expression. In addition to the electrical charge generated by each cell of our body, there is also a magnetic field that surrounds each cell, pictured in Figure 12. If you were to view the diagram in three dimensions, the shaded area would come to you vertically from the page. The human body as a whole exhibits a composite magnetic field, the sum of each individual field from each individual organ, tissue or bone cell. Electromagnetic cells within electromagnetic beings. We live the convergence two distinct though interrelated fields that determine, to a large extent, how we perceive ourselves, our world and how we function within those perceptions.

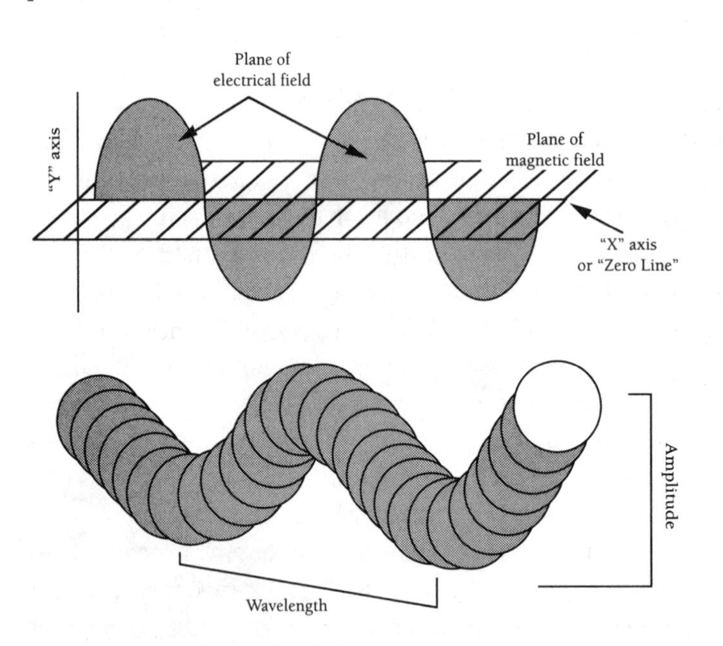

Figure 12
Schematic diagram illustrating the 90 degree offset between electrical and magnetic fields. The lower drawing is offered in illustration of the 3-D nature of the electrical field.

The electrical portion of your body is you, in your purest form as information, energy and light. This is your seed-core essence. This is you with no judgment, ego, fear or preconceived ideas regarding yourself, others or the world around you. The electrical aspect of you is what would historically be termed your *soul*. This aspect of you is not bound by dimensionality, planet or star. It is your soul that has traveled from a multitude of energetic systems to experience Earth in this cycle of consciousness. It is the soul essence that will eventually leave the Earth experience, at some chosen point in time, carrying the vibrational benefit of your Earth lives on toward a new experience.

The magnetic fields surrounding each cell of your body may be thought of as a buffer stabilizing the information of the soul within each cell. This buffer creates "drag" or friction around each cell, effectively interfering with your ability to fully access that body of information. Earth's fields of magnetics have, historically, been your safety zone between thought and manifestation. Early in this cycle of consciousness, magnetics were high, providing a distancing between the formulation of a thought and the consequences of that thought. Our group-body-consciousness was relatively new. Together, in the presence of one another, we learned/remembered the power and consequence of thought. It was during this time that higher magnetics were desirable. Both then, as well as now, it would be very confusing to have each thought and passing fantasy manifest as a reality in your life. Planetary magnetics were relatively high, ensuring that to manifest something in this world, we had to be very clear and really choose, or desire that which was being envisioned. Only then, could the seed of that thought be sustained long enough to be pulled down through the matrix of creation, crystallizing into something "real" in your world and your life. Now, as the intensity of the fields decrease, the lag time between a thought, and the realization of that thought, is decreasing proportionally. Perhaps you have noticed how quickly you are able to manifest in your world.

Lower magnetic fields provide the opportunity for change through the rapid manifestation of thought and feeling.

Meditation, Thought and Prayer

Within this world of polarity, choice and free will, you have an opportunity to realize your full potential as a divine expression of your Creator, becoming a creator yourself in the process. A portion of the lesson is that of the power of thought. You are learning that your feelings, as well as your thoughts, are "something," not just pictures in the eye of your mind or sensations of your body. As you think, feel and experience emotion, you create. Is there ever a time that you are not feeling, thinking or experiencing emotion? You create within each second of your life.

Within each moment lives your opportunity
to affirm or discourage life in your body.

Perhaps without full awareness of the process, you are remembering to function within the context of an extremely sophisticated system of creating "seed thoughts," amassing energy around those seeds, and bringing together components of the energy necessary to "gel" the seeds into your reality. Even in the technology of present day, one hallmark of sophisticated, user-friendly systems is the unknown complexity behind the scenes, providing the apparent simplicity to the user. The user does not have to *know* the intricacies of the system—only the ability to use them as they were designed.

Similarly, you do not have to know or understand any of the workings of your system of energy. In your complexity of being lives the simplicity of your expression. You may simply "be" in the purity of your intention and access all of the complex systems of thought, feeling and emotion representing your technology. Herein lies the beauty of this lifetime. You may choose to know if you wish—or simply be in the presence of experience and the outcome can remain the same. The ability to create (manifest) is a skill available to each individual, a direct result of three components of thought.

Clarity

For a thought to become real, it must be a clear, concise and sustained pattern of energy. A thought that remains a fuzzy collection of constantly changing patterns is probably incapable of sustaining itself long enough to crystallize as a manifestation. In the event that it does, it may appear to you as a series of incongruent, possibly confusing situations, mirroring the blueprints which it is attempting to express.

Duration

While clarity is a key component, without the ability to sustain the clarity, the seed is not maintained for a time sufficient to develop as it descends into successively denser levels of experience. It is a copy of the blueprint that is passed down through the creation matrix, gathering the energy at each level to sustain it into the next and become real, visibly manifest in your world.

Desire

It is the energy of emotion, or desire, that sustains the thought-seed as it descends the creation matrix. At any point, the seed may dissipate if the sustaining energy (desire) is lost. As the magnetic fields of Earth weaken, the thought becomes increasingly potent and less energy is required to sustain the energy for a decreased period of time.

Exercise A

An indication as to your ability to clearly create within your mind and sustain that creation may be found through the following demonstration. Examine Figure 13 for several seconds. When you believe that you know the figure, close your eyes and recreate the figure in your mind's eye. How long are you able to sustain the image before another image, though it may be fleeting, interferes? Try this exercise with the added component of color. You will find that the exercise and your ability to hold the image becomes better with practice.

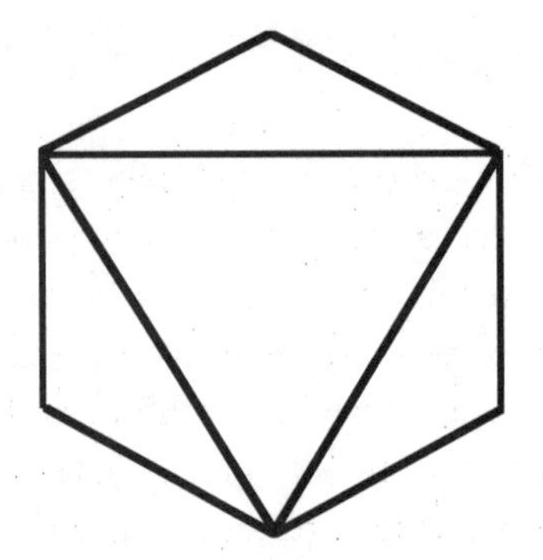

Figure 13 *Recreate this image in your mind's eye, making note of the image's duration before it is broken by another thought.*

Exercise B

Using Figure 14, repeat the previous exercise, first as a line drawing, then with color. Many will find that the second image appears too complex to accurately recreate. If they are able to do so, they may find that their mind wanders quickly, responding to external stimulus (street noise, household appliances, etc.).

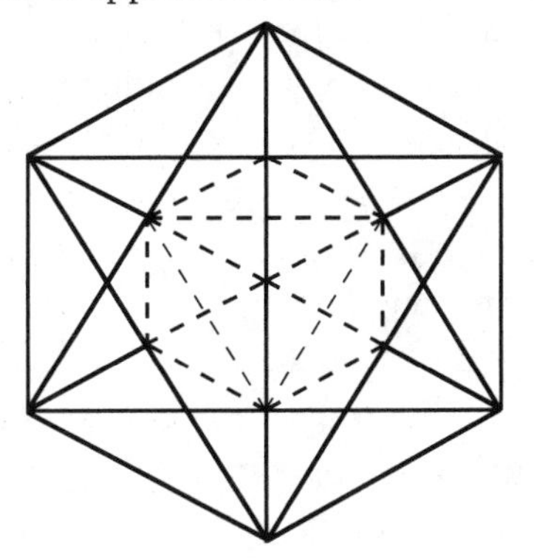

Figure 14 *Use this image of greater complexity to repeat Exercise A.*

Exercise C

Clearing your mind of previous images, focus your attention upon Figure 15.

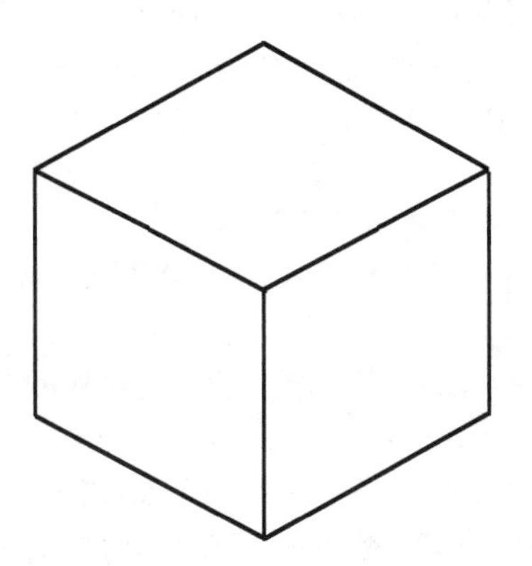

Figure 15 *Repeat exercises A and C using this image of moderate complexity.*

Repeat the previous exercise as a line drawing, making a mental note of your degree of success. You will probably find that this process is easier and more successful than that of Exercise B. There is an interesting relationship between Exercise A, B and C. The forms that you have imaged are representatives of a very special set of forms known as the Platonic Solids and are discussed in Chapter 2. The relationship is this: The form in Figure 15 is known as the Hexahedron, or cube, and is the parent to the previous forms. That is, the first two forms *live* within the third—if you have the Cube, by definition you have the previous forms of the Star Tetrahedron and Octahedron inherent within the form, though it may not be readily apparent.

The value of this process is in the demonstration of a point relative to creating and manifestation. In the creation of the Cube, *the other forms are already there.* They are implied. You did not have to specifically define them in your mind's eye. This very principle applies to you creating within your world. In imaging the desired outcome, allow creation to determine how the building blocks shall be drawn together. To define each building block individually may hinder your ability to create as constraints are placed upon the movement of the energy coagulating to form your creation.

Through

1. creating clarity of thought and feeling for your desired outcome while
2. sustaining the clarity long enough for creation to seed the blueprint into successively denser crystallizations

you will discover that the creation process requires little effort. You will also realize the consequences of your thought patterns very quickly. This is your invitation to become conscious of your thought and intent.

Meditation and *prayer* may each be considered as a form of thought, as each involves the use of directed intent, body circuits and a shift of body frequency through thought. Both meditation and prayer may be accomplished through specific techniques tailored to a specific purpose.

In Western cultures prayer is often considered as a form of requesting intervention. Through prayer, we are asking a power that we believe is beyond us to allow or prevent an event from unfolding. Examples of prayer commonly heard may include:

"Please God, let there be peace in the world."
"Please God, this one time, let my car make it to the gas station and I promise that I will never let my tank get this low again."
"Please God, allow me to become well and overcome this disease..."

In each of these instances, the individual initiating the prayer is asking for intervention from a higher power toward a situation over which he or she has been conditioned to accept a sense of powerlessness. Traditionally, Western cultures have been conditioned to view themselves as separate from creation, yet superior to that which surrounds them. In the illusion of that separation, they are unable to see, or identify with, the forces that drive creation, including the physical body. Feeling helpless to address events that unfold within their world and their lives, things for which they feel no responsibility appear to "just happen."

A fundamental concept developed within the ancient mystery schools, and shared in the traditions of indigenous peoples and many Eastern religions, is the idea that the human form is a *part of creation rather than separate from it.* As an integral part of creation, the individual plays an important role in the causes of events within creation. Though expressed within a vocabulary very different from the grid and matrix model of our conscious hologram, these ancient concepts are just as valid today as they were six thousand years ago.

Through the science of *mudra*, or the connecting body circuits (Chapter 3), the electrical resonator of our bodies is *linked* into the electrical circuitry of creation. Specific mudras accomplish a linkage with specific grids. As mentioned in Chapter 3, the most powerful mudra is that of the opened hands, finger tip to finger tip and palm chakra to palm chakra, used to access all grid zones possible throughout all of the creation matrix for an individual in a given moment. It is this mudra that we were taught when we want to pray, though we were seldom taught *why* we use this positioning of the hands and fingers.

Through the prayer position, you have positioned yourself to touch upon all zones of creation, each energetic reality that your body is capable of accessing, in that moment of no time and no space. Accessing the grids of information is only one step in the process. How do you then "tune" to the information contained within the grids?

Once the link is accomplished, it is through the quality of thoughts (clarity, duration and desire) that the circuitry of the body becomes *tuned* to the grids. You are always experiencing some thought or feeling; you are always tuned to something! The feelings that result from your life experience, your joy, pain and all variations of these extremes are, quite literally, teaching you how to think and feel frequency. Your thoughts are the tool that you use to bring yourself into resonance with various aspects of creation. Thoughts and feelings are your tuning mechanisms. When the tuning is optimum, you have established resonance, a connection of two-way exchange with a given level of reality.

Though you are, and have always been, a part of all that you see, at the instant of resonance you are aligned even closer to the energetic patterns to which your thoughts have carried you. In this resonance, or "oneness," you have the opportunity to plant and nurture the seeds of your creations. It is in the space of resonance, attained simply from your patterns of *thought coupled with feeling*, that you may direct energy most efficiently, consciously and with intent. It is in this space that you become the creator of your experience and may impact the events of your world while regulating the response of your body to that world. You intervene, on your own behalf, through the acknowledgment of your oneness with creation. The process that has allowed you to accomplish this tuning may be considered as a high form of mastery, using both directed thought and prayer, meshed into the process commonly known as meditation.

In reference to the prayers of intervention, it becomes apparent why meditation may be such an effective, *though passive*, form of intervention. Through the holographic model of consciousness, you are aware that the experience of each individual affects the whole to some degree. If this is truly clear to you, then it also becomes apparent why it may not be enough to simply pray for peace in the world, for example. Though well intentioned, the request for peace, in and of itself, is incomplete.

The highest form of intervention that an individual may offer within a given situation, is to move beyond the "asking" for something to be. The greatest expression of mastery toward a desired outcome may be found in *becoming* that which is desired. Herein may be one of the best guarded of the ancient secrets of personal mastery. You must become the very experiences that you most desire for yourself or others in life.

You must become the love, forgiveness, nonjudgment and Compassion that you choose in your relationships of friendship, romance and employment. If peace is the desired reality within the global matrix, then peace must become the reality within the local matrix of your presence. *You must become that peace.* It takes a relatively few number of "reference points" within the whole to bring about change in the expression of that whole.

SHIFTING FREQUENCY: THE SECOND KEY OF RESURRECTION

A second parameter of Earth's shift is that of *base resonant frequency,* or the base pulse of Earth. Within a given period of time, Earth pulses at a certain rhythm that may be measured as cycles or "beats" per second. These oscillations occur as a series of quick, rhythmic pulses, usually too rapid for our bodies to acknowledge. Our bodies do, however, detect these pulses through our brain centers and we are conditioned to average all of the signals together, perceiving them as a single continuous event. There are meditations specifically designed to bypass this conditioning, allowing the body to experience information as it is actually being perceived, short and subtle bursts of energy.

In recent years, the discovery of an atmospheric layer mirroring these zones of resonance is attributed to a German scientist named W. O. Schumann. From his work between 1952 and 1957, Schumann predicted the resonant frequencies that would come to bear his name. Nearly 53 years earlier, however, these frequencies had already been discovered and utilized by Nikola Tesla during research performed in 1899–1900 at his Colorado Springs facilities. Tesla's frequencies were based upon direct measurements of frequency measured, introduced and propagated through the Earth, rather than mirrored in the atmosphere. It is as a result of Tesla's work that we today have the polyphase system of generating and transporting energy that brings electricity into our homes, schools and businesses.

Both Tesla and Schumann found that Earth essentially functions as a massive planetary capacitor, storing and releasing electrical charge at specific intervals. These pulsations may be measured indirectly in our atmosphere within a zone known as the Ionosphere. A transient phenomenon, known as Ionospheric resonance, this layer of atmosphere responds to electrical "excitations" such as lightning, reflecting the frequencies that allow for an equal exchange of energy between Earth and atmosphere. Historically the points of resonance, both atmospheric and ground based, have been measured in cycles per second, or Hertz,* illustrated in the following sequence.

7.8 14 20 26 33 39 45

Range of historic Earth resonant frequencies, measured in cycles per second, or Hz.

* Hertz (Hz) A measure of the number of complete waves that pass a fixed point per unit time. For example, 10 complete waves passing a given point in one second would be referred to as 10 cycles per second or 10 Hertz. The unit is named after Heinrich Hertz, who discovered radio waves in 1886.

Measurements of these resonances may vary as much as plus or minus .5 Hertz per day. Our understandings of these frequencies and their relationship to our planet have several facets of interpretation. Among one school of researchers today, it is generally believed that

1. The Earth resonant frequencies measured in the atmosphere are not caused by anything internal to the Earth.
2. As long as the parameters of Earth's electromagnetic cavity remain unchanged, the Earth resonant frequencies remain unchanged.

The theories and research providing the basis for *Awakening to Zero Point* result from ancient beliefs and a growing realization in the scientific community that all systems of matter and energy, in fact, are interrelated. Life and life's patterns of energy coalesce and appear to be governed by hierarchical and systematically interrelated frequencies of wave form. Studies reported in *Science News,* January of 1997, suggest that even the formation of our universe may be dependent upon a geometric template resulting from frequency.[20] From this perspective, we must ask ourselves, "Is our Ionosphere separate, distinct and independent from Earth or part of our Earth/atmosphere system differing only in expression?"

Is the resonance measured in our atmosphere an independent phenomenon, as suggested, or does it reflect properties inherent within the Earth, displaying those properties when stimulated by a source such as lightning? Further, do the atmospheric readings accurately reflect Earth-based readings, as the atmosphere itself is undergoing a rapid change resulting from loss of planetary magnetics?

If our historic readings and interpretations of atmospheric resonance stem from an assumption that the readings change when the Ionosphere changes, our readings may not be saying what we have believed. *The Ionosphere, the source of atmospheric measurements, is changing rapidly.* The factor that allows this layer of atmosphere to exhibit properties of wave propagation is a function of Earth's magnetic fields! As the magnetics decline, the Ionosphere responds. In an article released in the Ham Radio journal, *WorldRadio,* we see that the Ionosphere's ability to hold and propagate frequency is declining and must be compensated for by boosting the signals of transmission.[21]

As the atmosphere-based readings fluctuate between 7.8 and 9.0 Hz, independent researchers have reported shifts in Earth-based frequencies toward a higher range of readings. The base frequency, which has hovered historically around 7.83 Hz, has been measured between 8.6, with reports of 9.0 Hz (unconfirmed by this author). Perhaps more importantly, higher harmonics along with a trend toward the phi ratio of .618 *between readings* are reported, indicating that the lower harmonics are complete and sustained. Recent research by the Institute of HeartMath and Dan Winter suggest that Earth's ability to sustain higher harmonics may be directly related to the electromagnetic web of conscious awareness overlaying our planet.

Regardless of the measurements, the net effect of Earth's frequency upon biological life includes the following:

All life within the envelope of Earth's vibratory influence attempts to "match" base frequencies to that of the Earth. To this end, each cell of your body is constantly shifting patterns of energy to achieve harmonic resonance to the reference signals of our planet.

In illustration of this point, please consider the following: When two electronic modules are placed near one another, with one vibrating quicker than the second, something interesting begins to happen. The module of lower frequency will have a *tendency* to match, through resonance, that of the higher pitch. This process is identical to that of the human energy system, which may be considered a module of composite frequencies reflecting individual cell and organ complexes. This module, our mind-spirit-body complex, when placed within the fields of another module (planetary or human) will have a *tendency* to move into resonance with the higher vibration. In the human energy system, however, the process takes on an additional component: the *willingness* of the conscious mind governing the body to adapt to the new range of vibration. The primary tool of adaptation is life itself, complete with the bundle of emotion, attitudes, perceptions, fears and beliefs that provide the framework for the challenges of life. One key factor in this process is the willingness of the individual to achieve balance within the energetic system using the tools of Choice and Free Will. These are the basic tools used to release old patterns of belief, lifestyle or relationships and adapt to new, more balanced patterns.

When accomplished successfully, this process is termed *healing* and the being is said to have *learned*. The ancients taught meditative techniques to bypass the logical mind in an effort to consciously feel these vibrations as well as sense the pulses of light. This is known as the difference between "seeing," or experiencing what is *actually* occurring, and "looking" or experiencing what is *expected* to occur. Within the last six years, measurements of Earth's fundamental vibration have been taken from independent sources throughout the country. Without knowledge of one another, each report that this pulsed frequency has accelerated gradually, with the greatest increase beginning in 1987. While this information remains to be confirmed in the open literature, it is precisely what the indigenous tribes have predicted for this time in history.

The net result of this increase is that each cell of our bodies seeks to match the rhythmic "heartbeat," or reference frequency, to that of Earth. Moving into the resonant pattern of a higher tone, each life form, including human, is attempting to map out a new rhythm, or "signature frequency." This frequency is indicative of the composite vibration of Earth at a given point in time. It would be possible to entrain the cellular heartbeat of the physical body immediately, if it were not for variables that provide resistance to the entrainment, such as belief systems. To the degree that an individual clings onto structures and beliefs that no longer fit into the new paradigm, to that degree will the individual encounter resistance to the entrainment of cellular frequency with planetary frequency.

What is the target value of Earth's shifting frequency of resonance? Where will the base resonant frequency of Earth be at the close of the cycle? Before this question may be answered, it is necessary to introduce the concept of a very special series of numbers—literally codes for fundamental aspects of life on Earth.

1	1	2	3	5	8	13
21	34	55	89	144	233	377
610	987	1597	2584	4181	6765	10946
		17711	28657	46368...		

Table 2 *The first 24 members of the Fibonacci series based on "1."*

Beginning with the integer 1 and adding 1 to itself, a pattern is developed as any one term results from the sum of the two previous terms. One plus one equals two. Two plus one equals three. Three plus two equals five and so on. Table 2 represents a portion of this essentially infinite series. Deceptively simple, this series of numbers is a fundamental code of the patterns of life on Earth. An example of parameters that are governed by this series includes:

- Relative proportions of the human body
- Ratios of males to females in an uncontrolled population
- Branching patterns of trees, plants, shrubs (without pruning)
- Dendritic branching patterns of lightning and root systems

The actual ratio may be determined in the division of any term in this series by the next higher term, producing an interesting phenomenon.

$$21 / 34 = .6176 \qquad 89 / 144 = .6180$$

Though the terms vary, the ratio of the terms is relatively constant! Used in ancient Greece to describe the most pleasing proportions of human anatomy, as well as classical Greek architecture, the ratio is known today as the *Golden Mean* or the *Golden Ratio*; the fractional value of .618. This code is so fundamental to "life" on Earth that it is not surprising to find it also governing the fundamental vibration of the planet.

Historically Earth has pulsed at approximately 8 cycles per second, 7.8 until recently. This value has been so stable that global military communications have developed relying upon this "constant." As of this writing, the oscillations are increasing, having moved into the range of 8.6 cycles per second. The target frequency of Earth resonance is the next member in the sequence of values that govern this parameter: 13 cycles per second. It is 13 cycles per second that may become our new base resonant frequency with all harmonics based upon integer multiples of this fundamental vibration. This is the frequency that may trigger resonance with the new grid-matrix complex, signaling the close of the present cycle of evolution and the beginning of the "New Age." If ancient timetables are correct, it will happen within our lifetime.

1, 1, 2, 3, 5, **8, → 13,** 21...

Table 3 *Historically, Earth has hovered around a base resonant frequency of 8 Hz. Now, at the close of the cycle, this fundamental vibration is moving toward a Fibonacci target frequency of 13 Hz.*

THE COLLECTIVE INITIATION

Human experience is, and has always been, intimately tied to the strength, or density, of the magnetic fields surrounding Earth. The decrease in planetary magnetics seen today has a profound effect upon all life. Many remains of the ancient technologies, when viewed today, indicate a preoccupation with the building of specialized chambers: tuned resonant cavities of experience. Frequently, these structures of precise geometry provide an environment lessening the effects of planetary magnetics within the chamber during use. Perhaps the best example of this technology may be seen in the Great Pyramid of Giza, where the upper known chambers have significantly lower readings of Earth magnetics than the lower chambers.

The net effect of this environment is that those who introduce themselves into these null-magnetic fields have nearly direct access to themselves as pure information without the interference patterns of judgment, fear, ego, or the buffers of "lag time" between thought, and the consequence of a thought. With that access comes the opportunity of cellular and core-level change, *directed through the will of the individual*, consciously toward the life parameters of choice. As the individual, *the ancient initiate*, chose to think and feel differently, he or she was able to consciously alter the frequency of each cell of the mind, body and spirit complex. This shift allowed the fullest expression of being, available within that moment. This process was demonstrated for us today, simulated within the chambers of ancient temple and initiation sites during a time in which those same parameters were not available elsewhere.

Though not limited to Egypt, the sequence of initiation into immortality (Christhood) demonstrated within the temples of this sacred land, will help to demonstrate the conscious path of the initiate then, as well as application to events occurring within your life now. For over 60 centuries the magnificent temple complexes situated along the Nile River have provided the fuel for controversy between scholars, historians and even the Egyptians themselves, as to who created the monoliths, how they were constructed and for what purpose. Traditional historians relate temples to "gods" that they believe were immortalized as statues, reliefs and hieroglyphics specific to each complex. The initiates themselves tell a different story; a story of the visions and goals of life, not unlike the visions and goals of each individual today.

It is true that each temple site had a primary deity personifying the focus of that particular temple. Depicted as monumental statues and in deep relief, both inside and outside of the walls, in most instances these figures were not intended to represent external gods. Rather, each temple was dedicated to the isolation and focus of one component of the human psyche, one unique aspect of personality that must be balanced prior to moving into higher levels of mastery. Through isolating each component and consciously working toward the compassionate balance of that trait, the initiates were able to address those fragile portions of themselves that were "lost" through distortions occurring within the course of their lives.

For example, on the perimeter of the Luxor temple complex on the east bank of the Nile lies a smaller, less known chamber, the Temple of Sekmet. Represented as the body of a woman with the head of a lioness, scholars relate Sekmet to the act and science of "war." Wherever Sekmet is seen, the common belief is that she was placed at that particular location to protect from war, to prepare for war or to bring good fortune to those engaged in warfare. To the initiates using this chamber, both then and now however, Sekmet represents something much more powerful on a much deeper and more personalized level.

To you as an initiate, Sekmet provides a focus for an aspect of personality that lives within you, the "warrior within" that you are asked to call upon at some time during your life to "break the barriers" and "tear down the walls" of resistance blocking the path of your life's vision. These barriers may not always be physical obstructions. There are instances where the control, fear, judgment or ego, the will of another individual or system became an obstacle to the completion of a choice or commitment made within your life. It is the energy of Sekmet, the warrior that lives within you, that is called upon to overcome those barriers. There is both an art and a science in the knowing of when and how to call upon this awesome force, *as well as the wisdom to know when it is no longer appropriate and when it no longer serves you.*

Initiates studying within this Temple would subject themselves, time after time, to situations that would call upon their knowledge of the Sekmet energy, as well as their developing wisdom as to when the warrior was not required in a given situation. Throughout your path of initiation, *your life*, you have been asked to call upon the *knowledge* of the Sekmet within you to break down the walls of resistance that you have encountered within or drawn into your life. Through an iterative process, you are also developing the *wisdom* to discern when the warrior aspect of yourself is not an appropriate response to a given situation. This is the wisdom that may only be attained through your unique experience and perspectives of life.

The Temple of Kom Ombo, the Temple of Light and Darkness, is the only Temple in Egypt dedicated to two deities simultaneously. Looking into Kom Ombo from the main entrance the Temple is divided in half. The right half is dedicated to the darkness represented by Sobek, the body of a man and the head of a crocodile. There are chambers in this portion of the complex that still house the mummified remains of crocodiles used in ceremony. The left half dedicated to the light, symbolized as the human body hawk-headed image of Horus, the only son resulting from the union of Isis and Osiris. In this Temple the initiate would be asked to find the balance between the polarities of darkness and light, and once knowing the balance, were asked to live that balance within their lives.

To demonstrate their willingness and ability to live these principles, each initiate would be required to complete a test within the Temple walls. Toward the rear of the Temple, situated along the midpoint between dark and light, lies the remains of an initiation chamber used in the demonstration. Configured as a long, narrow, rectangular passageway, one end of the chamber was elevated approximately 45 degrees to a height of ten feet above the Temple floor. The inside of the passageway itself was relatively smooth with corners angled at 90 degrees and a uniform surface throughout. About two thirds of the way through this tunnel, however, an obstacle is formed as an elongated portion of the ceiling, effectively blocking the passage except for a narrow space beneath the obstacle. On occasion, the entire tunnel system could be filled with water, the angle effectively preventing daylight from reaching into the tunnel.

Initiates were required to enter the passageway at the top, above the Temple floor, and swim through the long tunnel, emerging on the other side not knowing what lay between the entrance and exits of this chamber. With a single breath they began their opportunity to demonstrate their mastery. By this stage in the initiation process, the intuitive skills and use of all senses became imperative in the sensing of direction and obstacles within the tunnel. The fear of running out of breath, of itself, caused enough anxiety in some to prevent them from wisely using that breath. In their fear, they would perish. Others, moving quickly in an effort to conserve breath, would experience a powerful collision with the barrier. The jolt of impact could cause them to expel their breath and in the ensuing panic, fail to locate the opening below the barrier. If the impact were severe enough, the initiate would lose consciousness and drown. To those who were able to approach this initiation using all of the tools of consciousness available to them, trusting in themselves and the process of their lives, they would emerge from the other side of the tunnel, in light, signifying the birth of a new phase in their lives; the ability to successfully balance the energy of dark and light. In the completion of the tunnel initiation, initiates would have demonstrated to themselves that they could trust in themselves and rely upon their intuitive abilities. In that trust they had learned not to give their power (or their lives) away to fear.

In a fashion similar to the ancient initiates,
you are being asked to live the initiation of light and dark many times
throughout each day of your life.

In each choice that you make, regardless of how insignificant it may seem to you at the time, your are redefining your relationship within the polarities available to you in this experience: the pole of darkness and the pole of light. The words through which you express yourself, the manner in which you treat others through your daily exchanges, the manner in which you treat yourself, the kinds of food that you choose, the kinds of information that you choose to immerse yourself in, each of these is an expression of the manner in which you view yourself relative to light and dark. It is important at this point to reiterate, that in any of these choices, there is no "right" and no "wrong" (bearing in mind that by living within a given society you have "agreed" to adhere to the guidelines of that society). There is only choice, the consequence of choice and the resulting feeling from the choice. These *feelings* are the root of issues being addressed on a global scale today. How do we feel, as individuals and a group consciousness, about what we do or have done? Choice knows of no right or wrong until it is judged by comparing to another standard of choice. Allowing yourself, and others, to experience the consequences of choice without judgment is one of our highest levels of life mastery.

Additional temples and chambers provided similar environments: Saqqara, for example, and the Temples of sound and vibration; Dendera and the Temples of love, nurturing and healing; the Osirion complex and the Temple of Resurrection. Each temple identified and isolated one aspect of personality seeking balance (mastery), before moving on to the higher levels of initiation. All mastery was in preparation for *something even greater,* a sometimes nebulous and elusive stage of expression aspired to by many and attained by few. This level of mastery allowed for the conscious vibration of each cell within the mind, body and spirit complex to a new level of expression. In effect this level of mastery was mirroring the process of a conscious dimensional shift.

In viewing life within the context of the lesson that each experience brings, it becomes apparent that you are attracting opportunities that allow you to examine your personality, in a fashion similar to the ancient initiates. The primary difference between you and the initiate six thousand years ago is that perhaps no one has reminded you that your life is an initiation. Perhaps no one has reminded you that you are an initiate of the highest order! Though you may have felt intuitively that you have been preparing for something all of your life, training and wondering, possibly no one has said to you what that *something* is. Each experience, each encounter with another, every relationship, regardless of duration, everything that you have chosen to experience is preparing you, emotionally, physically and psychologically, for the monumental change that you will experience within this lifetime. You are living now the processes modeled previously in the chambers of initiation for the last 11,000 years. Your life is preparing you for The Shift of the Ages. To the degree that you are able to love without fear, experience without judgment and allow through Compassion, to that degree are you preparing yourself to survive The Shift emotionally and psychologically, as well as physically. You are already living the initiation!

The beauty of the initiation is inherent within the process itself. You do not have to know anything about magnetics, frequency, grids, matrices, dimensionality or Sacred Geometry. In the complexity of your being you may simply "be" in all of the truth, purity and wholeness that you are capable of within a given moment. Through living that purity, you may attain the same mastery as those choosing to "know" and "engineer" their bodies and experience. Both paths are valid. Both will eventually lead to the "One."

The temple initiates would spend years, perhaps an entire lifetime, addressing each aspect of personality. As in our lives today, some would become "stuck" in one, or some combination of personality, forgetting that they were not their experience. They would begin to believe that they were all of the pain, all of the joy, the successes and the failures. They began to identify so strongly with their experience that they believed the experience was them. Not unlike the initiations of life today, there are some who have remembered the lessons of the life experience as their passing; they have died remembering. For those that did remember, their final opportunity to demonstrate mastery involved a process not unlike that which will be experienced by each human remaining upon Earth during the days of the close of this conscious cycle. Those living the process of Earth, mind, body and spirit allowing a higher octave of expression will experience the opportunity of a phase transition known as Resurrection.

At some point during the life of the initiate, he/she consciously chose to begin a process during which they would model within a chamber at that time, a series of events that Earth is living now. The process would begin within the lower chamber of the structure known today as the Pyramid of Cheops or the Great Pyramid (Fig. 16).

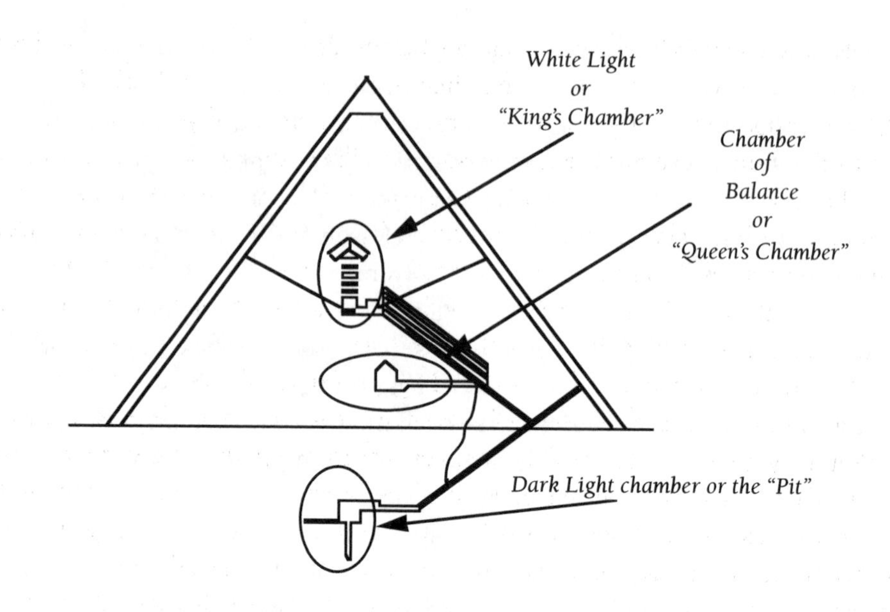

Figure 16 *A schematic cross section of the Great Pyramid illustrating the chambers relevant to the Collective Initiation. This drawing is not to scale. The ancient texts tell us that there are in excess of 30 chambers above ground level within the pyramid, predicted to be found before the close of this decade.*

By virtue of the passive dynamics inherent in the building of this structure, the initiate in the lower chamber was completely immersed within the energies of pure "Dark Light." Silently, in the complete absence of visible light, the opportunity of mastery would begin as the opportunity to realize and address issues that had not found the resolution of healing. The issues were based in the illusion of the individual's own worst fears. They would be different for each.

It is in this chamber that the value of years of experience within the Nile temples became apparent. *It will be during this portion of your life that all of your training, within the temples of your relationships, becomes apparent.* In the mastery of individual aspects of personality, the tools are available to examine the fear for what it is, and not what the life experience makes it to be. If the initiate had not learned to embrace fear as a powerful ally in the knowledge of self, in that moment the fear would become reality, seeking resolution as demonstration of a high level of mastery.

There were initiates who demonstrated their mastery of this experience through separating from their body in death. At last, in the depths of the chamber allowing the illusion of their own greatest fear,* they would see through their fears by experiencing their worst possible outcome, seeing themselves through this outcome. This metaphor of their initiation is not unlike what is/has/will happen to you at some point in your life. Having attained a degree of personal mastery through your life experience,

* As late as 1987, a specialized portion of this chamber leading to the legendary Halls of Amenti was closed to the public, as individuals would crawl into the chamber and, for "unknown reasons," not survive. (They were living their worst fears, unconsciously.) Today, 1997, you may descend into the lower chamber but will be prevented from moving into the initiation chamber by a steel grating.

you will allow yourself to be drawn into a situation or circumstance which will represent, to only you, your worst fear. This time is referred to as the "Dark Night of the Soul," your opportunity of greatest mastery. In that moment of your life, you will be asked to draw upon every particle of wisdom available to you, from deep within, to negate the apparent power that you have given to your fear. Often, this becomes an iterative process with the effect of the fear becoming less and less in each iteration. The goal of the process is to arrive at the realization that no matter what comes your way, no matter what you are faced with, no matter how devastating a situation may appear, you are safe and intact. There can be no fear if the fear is given no power upon which to feed.

Ancient texts remind us that a Dark Night of the Soul may only occur after we have drawn to ourselves all of the tools necessary to see our way through the experience. You may go for years, possibly lifetimes, without experiencing your Dark Night if that is what it takes for you to gather all of the experiences necessary to masterfully embrace your experience. Your worst fears of being alone, left, without money or in poor health, for example, may only occur when you have had the experiences that have shown you how to survive these experiences intact and with grace.

Following survival of the Dark Night of the Soul, initiates would move into the middle or "Queen's" chamber to demonstrate their ability to balance the polarities of Light and Dark. The Nile Temple initiations (Kom Ombo) had prepared them to demonstrate this level of mastery. In this particular chamber, they would be required to "hold the balance" for an unspecified period of time.

Having just emerged from the lower chamber experience of pure dark light and the resolution of the initiate's worst fears, the challenge of this chamber was to resist the urge to overcompensate by moving exclusively into the light. This chamber models for each of us a path to be aware of. Having survived a dark light experience, the tendency is to overcompensate and move into the opposite polarity.

The overcompensation may not be effective because the initiate, both ancient and modern, is still living in a polarity experience and must embrace both experiences without judgment, in demonstration of a high level of mastery. Both polarities must be experienced however, known and accepted without bias, before the "charge" may be balanced and mastery attained.

Upon completion of the demonstration of balance in the chamber of light and dark, the Queen's Chamber, the initiate would advance into the uppermost of the known chambers, the White Light or "King's" Chamber. It is within this chamber that the final stage of initiation would be demonstrated. Once completed successfully, the initiate would no longer be subject to death, illness, disease or the limitations of linear time. In the completion of this level of the mastery over fear, emotion, and the illusion of separation between body, mind and spirit, the individual had attained the ability to experience a higher octave of expression; that of a Resurrected being.

The King's Chamber of the Great Pyramid offers a very unique environment, the source of which remains a mystery to modern science. Within the tuned resonant cavity of the Chamber of Light, measurements of Earth's magnetics drop to nearly zero. In historic times, the area inside of this chamber provided an environment that was unavailable outside of the cavity. Today, Earth is moving toward the identical environment that was modeled in this chamber over 10,000 years ago, that of low magnetics.

Earlier in this text, it was noted that in the presence of a relatively strong field of magnetics, a lag time is produced between thought and the crystallization of that thought. Additionally, it is within the buffers of dense magnetic fields that the interference patterns of emotion are stored holographically surrounding, as well as within, the body. The absence of magnetics provides the opportunity for direct access to the individual as pure information. It is within this environment, without the safety net of magnetics, that thought becomes very potent. Through the gift of this very pure environment, initiates, having mastered personality, ego, fear, judgment, forgiveness and Compassion simply begin to think. Possibly for the first time in life, the thoughts are their own and not the imprints of a consensus reality, or the constraints imposed upon them through earlier conditioning. Through direct access to themselves, the initiates become "real." Within that reality, they carry themselves, beyond the limitations resulting from the life experience. The difference in thought is a vibration, one which entrains all other vibrations of the body. Through thought, in a zero magnetic environment, the initiate has the opportunity to unplug from the bonds of this reality and simultaneously move into a higher expression of experience. The conscious shift of each cell of the mind, spirit and body into a higher octave of expression is the act of Resurrection.

The process is accentuated within the White Light Chamber through the precise tuning of the chamber to the human body. The proportions of the chamber itself are based on fundamental proportions found within the Platonic Forms and mirrored through the "codes" of the human body. Within the chamber itself is an additional chamber, sometimes referred to as the sarcophagus. Formed from a single block of beautiful rose quartz granite, the interior dimensions of this container are identical to the exterior dimensions of the Biblical Ark of the Covenant. Though damaged by tourists over the years, the sarcophagus remains today, tuned to a perfect "A." This may be heard on the first track of the flute solos recorded as *Inside the Great Pyramid* by Paul Horn. It is within this rose granite enclosure that the initiate would lie, with the matching granite lid sealing them off from the outside world for a period of three days, or 72 hours. This is the identical duration of the initiation modeled to us by Christ Jesus through the Crucifixion/Resurrection sequence.

Additionally three days is the prophesied duration of the "in between" time at the close of our cycle, just prior to the time that Earth begins to generate new fields of magnetics; the dawn of the New Age. During the three days, in an environment of increased frequency and decreased magnetics, the initiates would consciously vibrate themselves into resonance within the grid-matrix framework of higher octave knowledge to access that knowledge, touching the Christed grids of a fourth dimensional state-space. Through the reintegration within the sarcophagus they would demonstrate their newfound wisdom to others who had completed the same process. In the demonstration they would then show that they had, indeed, remembered. If the initiation was completed successfully the initiate became immortal, living without sickness, disease and deterioration until the choice was made to ascend permanently into the experience to which resonance had been established through the Resurrection process.

Within the pyramid initiation, it becomes apparent why each individual would choose to balance all aspects of personality prior to moving into the Chamber of dark light. In the presence of pure dark light, thought patterns arising from the negative aspect of the polarity would be accentuated, as an indication of hidden or unconscious fears to be addressed. Equally desirable was the resolution of those fears, or any other patterns originating within the negative aspect of the positive/negative polarity.

In the presence of pure white light, in an environment of increased frequency coupled with decreasing magnetics, any patterns of thought may manifest instantaneously. If the patterns are those of Compassion, forgiveness and love, those patterns become the experience. If the patterns are those of residual doubt, mistrust, anger and fear, those patterns become the experience, as well as the reality. The process within the white light chamber was simulating at that time, an experience identical to that of Earth at this time; an environment of low magnetics and high-frequency. Today, you are living a process identical to that of each initiate that has ever pierced the veils of life's mystery.

Earth is essentially crystalline in nature with one of the primary components that of silica, or quartz (SiO_2). Life upon Earth is a mirror to the process of being surrounded by a grid of Earth resonant stones. All matter in general, and each human cell specifically, strives to maintain a constant resonance, or "tuning," with the pulsed heartbeat of Earth. As our physical bodies attempt to maintain this attunement, those areas holding inharmonic frequencies of experiences "light up" making themselves known so that they may be addressed and healed (balanced). Cells and organs holding the energy of trauma (psychological, emotional or physical) begin to vibrate out of "sync," dislodging the energy of emotions that no longer fit, allowing them to seek balance by relocating themselves to a point within the body where they may be stored in a balanced fashion. When this is accomplished it is said that the individual has "remembered."

Ancient techniques of electrical and magnetic balance, known as *crystalline healing*, provide an excellent metaphor to describe the interaction between Earth and the human body occurring today. In the science-art of crystalline healing, the client is surrounded by Earth resonant stones effectively forming a grid pattern around, and upon, the body. To aid in clarification, the modern science of electrophysics describes one possibility in the interaction between two electromagnetic modules (Earth and human) as follows:

> If one of the modules vibrates at a higher frequency relative to the second, there is a tendency for the module of lower frequency to become entrained, or "tuned," so that it matches the higher frequency module.

Each cell within each organ, and the body overall, attempt to match the higher reference frequency offered by the stones. Those cells located within the body where patterns of experience (emotion) have been resolved, achieve resonance easily. Those cells storing inharmonic patterns, that have not found resolution, cannot achieve resonance so easily and "light up" as memories, emotion, sensations and feeling. Some memories are so dramatic that the individual actually somatizes the experience, lives the old patterns physically in the present moment. As an initiate living upon the crystalline sphere of Earth (approximately 98% quartz as silica), you are living the "crystal healing" of Earth's shift, attempting to match resonance with the reference frequency of our planet.

The time of The Shift is not about death and endings. Rather, The Shift mirrors our choice of life. The Shift of the Ages is a new expression of life and the birthing of a new, integrated wisdom. The opportunity of Earth is the opportunity of experience, whatever that may mean to you. Your life is your greatest tool of mastery. You will discover that fear, *your fear*, is your greatest ally in preparation for the experience. Through discovering your greatest fears, you will be led to your greatest healings in your choice to remember your truest nature. Please, I invite you to move beyond your conditionings of fear! Ancient initiates addressed the path consciously, knowing the reason and purpose of each event and experience. You have been asked to do the same, possibly without the benefit of knowing why. Through the act of simply living your life, you are literally living the processes that were modeled, predicted and prophesied over 2,000 years ago. You are the initiate. You are living the Collective Initiation.

SIMULTANEOUS REALITIES: THE MAGNETIC/FREQUENCY RELATIONSHIP

The significance of the marked increase in base resonant frequency now, within the last ten years of the cycle, is critical in the understanding of the Christing process. By definition, Resurrection is the conscious increase of body frequency, the entire mind/body/spirit complex moving into resonance with new information. This variability of frequency, then, becomes the definition of multidimensional realities.

Multidimensional realities may be defined as multiple events/experiences occurring simultaneously, occupying the same space at the same time, within different zones of frequency. Each event is taking place within its own envelope of expression, usually unaware of other events occurring within the same continuum.

For example, the reality that you are experiencing at this moment is probably that of what is known as 3-space, the familiar third dimension. If you were to vibrate each cell of your body, mind and spirit complex (either mechanically or through feeling) into a specific harmonic (integer multiple) of 4-space, you would no longer be engaged within the experience of a third dimensional reality. Those experiencing you visually in 3-space would no longer be able to detect you, though you had not changed geographic location! You would have vibrated beyond their range of perception into a simultaneous range of existence "keyed" to a higher dimensional state-space.

A reference to higher dimensions is not a reference to experiences above the point of your physical location. You, through the polarized interference pattern of your body, are the sum of all dimensional experience, focused through you, at your particular "wrinkle" in the matrix. Quite literally, within you exist all possibilities, as potential; you are the beginning and the end, the Alpha and the Omega, the first and the last. It is through the intentional "stressing" of our time-space fabric as increased body frequency that alternate realities are created and experienced.

The decrease of planetary magnetic fields creates the opportunity for this very experience, through bypassing the buffers of distorted emotion (fear and judgment). Techniques to accomplish the bypass, from within, were offered through the ancient mystery schools and have remained to present day, passed down through code, parable or in an unwritten format to those seeking the path of Resurrection.

For all experiencing Earth at present, the beauty of the process is apparent. As the cycle draws to a close, planetary magnetics are decreasing rapidly to allow direct access to consciousness as information—without lag time buffering the experience. You are beginning to experience nearly instantaneous consequences of your thoughts. At the same time, each cell of the mind, body and spirit complex is striving to maintain resonance, with the Earth—a resonance that is moving through an unprecedented shift into a higher band of frequency. The Shift has already begun and you are participating in the process!

PROBABLE CHRONOLOGY OF THE SHIFT: THE SEQUENCES

From the two parameters of magnetics and frequency alone, it becomes possible to view the process of The Shift from a new perspective. Without fear, judgment, ego or the dogma of religious beliefs, the process of The Shift is moving toward a physical, digitally measurable goal; that of low magnetics and high base resonant frequency. As life forms within the protective envelope of Earth, each individual is subject to the same transitional parameters as the planet itself. The processes leading up to The Shift are already under way. Composite readings of planetary magnetics are approximately 50% less than those in samples from 1,500 years ago, with measurements currently near 8.0×10^{25} gauss units.[22] Through the lessening of magnetic interference patterns, individuals continue to have greater access to themselves as information, without the buffers of magnetic interference.

At the same time, base resonant frequencies continue to increase. Each cell in the physical body is trying to match the reference that Earth is providing by vibrating at a higher rate. Each aspect of the body-mind matrix must arrive at this alignment eventually; this is the balance of peace, well-being and union that is historically the goal of the spiritual disciplines.

Preshift dynamics may become more pronounced for some unspecified period of time. These are finite processes, however, and not necessarily linear in nature. It is not possible to predict how the fields will vary based on a straight-line extrapolation referenced from a measurement of today. Stanford University geophysicist Allan Cox, for example, estimated that at the rate of magnetic decay observed in the early 1970s, planetary magnetic fields would reach zero in approximately 2,000 years.[23] The assumption at the time of the studies was that the fields were decaying *at a constant rate*. In recent years, observations indicate that the decay appears to be geometric in nature, moving into more of an extreme as each set of values drifts away from historic readings.

Historically, magnetic fields appear to have reached some threshold and declined very quickly prior to magnetic reversals, as evidenced by the frozen "fossils" of mammals found in the Arctic ice of Northern Siberia. Paleontological evidence indicates that these warm-blooded animals appear to have been in a semitropical environment one instant and subjected to an Arctic freeze, *within seconds*, the tropical vegetation in their mouths and stomachs still viable as a food source after 10,000 years. Ice cores and fossilized (frozen) remains in the polar areas seem to support this idea as do Biblical accounts of the event.

> *Lo! I tell you a mystery. We shall not all sleep, but we shall all be changed, in a moment, in the twinkling of an eye...*
>
> THE HOLY BIBLE, I CORINTHIANS 15:51

If the geologic record of previous polar reversals is accurate, the indications are that the magnetic fields of Earth will continue to decrease. There are no accurate records of Earth's fundamental frequencies prior to 1899–1900, making extrapolations of Earth frequency from past data difficult. The present indicators point to the continued increase of this fundamental parameter toward a key resonance. If our past provides a model for our present, upon reaching a critical threshold, magnetics may bottom out at or near zero, and frequency will reach a new value of harmonic resonance, possibly 13 cycles per second. This point completes the *process* and begins the *event* that has been the subject of religious traditions and beliefs for thousands of years.

Earth may experience a period of unprecedented quiet. During this time of decreasing motion, the rotation of the inner and outer cores of Earth has dropped to near zero. Though the slowing of the rotation is gradual, it will probably not be "clean" in the sense that the plastic/molten material will continue to shift, as a sloshing motion, creating erratic fields of magnetics for the duration of the motion. (For example, holding a glass of water in your hand and slowing down as you walk across a room, the water continues to slosh even though you have come to a stop.) It is this slowing of interlocking fields of energy, related through the "coupling effect" of core/mantle/crust/atmosphere, that is producing unusual patterns of energy, affecting the lives and commerce of nearly every nation. Expressed as records of temperature, wind, rainfall and increased seismic activity, along both old and new plate boundaries, it becomes possible to see why the patterns have developed and why they will continue until the planet's "goal" of balanced patterns of energy is achieved. Earth will continue to have gravity (an electromagnetic function) throughout the process, unaffected by the weakened fields of magnetics.

At the point of threshold resonance, at or near 13 Hz, magnetics are approaching zero. The chaos leading up to this time has ceased. The dynamo "driving" the motion of the 20(+) tectonic plates that comprise the planet's surface as well as atmospheric currents are still. There are no active weather patterns. That portion of Earth which is bathed in sunlight will remain in sunlight for the duration of the process—that which is in darkness will remain in darkness for the same period of time. (An exception to this scenario may occur if the Hopi prophecy of a "new sun" (the blue star) in our solar system materializes. These has been speculation that the impact of a foreign body or other phenomenon may ignite Jupiter's pre-sun atmosphere. In this instance, Earth will be bathed in light from all directions, though the quality of light will vary. As the prophecies indicate, Earth will never again see darkness. During the time that Earth experiences these events, we will experience the "Zero Point" of this cycle of awareness. This is the dawn of a New Age, "The Shift of the Ages."

An analogy of a common technology may help to clarify the concepts of the breaking down of conscious grids during The Shift. Many today use a personal computer for home or business applications and are familiar with the basic components and operation of this device. The component of primary interest here is the "memory" of the computer. When you leave the office for the evening and switch off the power to the machine, you feel that you have "shut down" the device by cutting off the flow of electricity. When you return and "power up," if all goes well, the operating system appears and you are able to begin your work. The question is this:

If the power to the machine was, in fact, shut down, how was the information of the operating system retained within the memory of the computer?

The answer to this question is a very good metaphor to the relationship between conscious memory and the magnetic fields of Earth.

Within each microcomputer there is a small battery that is loaded into the unit, normally at the time of manufacture. Following power down, this battery continues to provide a trickle charge, a small amount of electrical current that holds the magnetic alignment of information in place until the system is powered up for use. On occasion, normally after about four years of operation, the battery will fail to hold a charge and one morning, as the power is turned on, the system will fail to initialize. *It has lost the magnetic alignment of its information and must now be reformatted and reloaded.* The old system is wiped clean to create a "blank slate" upon which to load a new system from an external source.

This analogy provides a working model toward the understanding of conscious memory within the stabilizing fields of Earth's magnetics. The rotation of Earth provides a static electrical charge, analogous to the battery, holding the conscious information of the grids of man in alignment within the magnetic environment. A decrease in the rate of Earth's rotation produces a decrease in the density of the magnetic fields as well as a decrease in the static or trickle charge stabilizing the orientation of memory patterns. Both of these phenomena are occurring at present, though the relationship between the two may not be apparent.

The anomalies of magnetics and frequency are occurring at present (late 1994) and are viewed as discrete, separate and unrelated events by most individuals. All are part of something much more significant than any single event. *The key to witnessing the possibilities of our future is to bear in mind that as our Earth changes, we change as well.*

We are not the same beings that we were even five short years ago. We will not be the same beings witnessing the moment of The Shift as we are today. For this very reason, the physical consequences will not have the same meaning to us then as they do now. As we move into what the ancients called the "new dream," our link to temperature, weather, magnetics and cycles of light and dark are redefined. I believe that the processes unfolding within our lives are healthy, natural and our way of creating a new way of life. We happen to be the generation that is living the interface of the new as its template is gracefully superimposed upon all that we have ever known or held as true. Us, living in this time of transition, is where we have the greatest opportunity to demonstrate our strength, wisdom and Compassion as we live our mastery of two simultaneous worlds in the presence of one another! From these perspectives, I invite you to entertain the following, allowing for all possibility of change within us, we who have come to this world at this time to know ourselves in all ways.

The complete reversal of Earth's magnetic fields may well include the following physical effects:

- Earth's healing. Materials and compounds that are born of and affirm life, experience graceful transition. Frequency "filters" ensure the removal of toxic and inharmonic compounds, as well as the thought process that has allowed them within the conscious matrix. Products of anything other than Earth resonant materials (substances that have been "forced" into unnatural combinations under artificial conditions) may disassociate as the parameters allowing the substances no longer exist.

- The temporary failure of electrical and magnetic technologies based upon the movement of electrons through a conductor. "Modern" technologies have been developed under the premise that electrons move at a "fixed" speed and that the media carrying the electrons will not change. Internal combustion engines, conventional power systems, computers, and many broadcast technologies may prove to be temporarily invalid as the parameters that allow these devices to function are no longer present. Following the polar reversal, technologies that depend upon a "directional" flow of electrons may prove invalid as the flow of electrons is reversed.

- The simultaneous overlay of two conscious and spiritual cycles of human evolution. In the words of the Hopi, "When the heart of man and the mind of man become so distant that they are no longer one, Earth heals herself through the catastrophic events of change."[24]

- Temporary equilibrium or balance of the dynamic forces driving the cycle. The Shift results in Earth moving into resonance with the "light" or the zone of intense and focused frequency that has initiated the process and driven it to balance.

THE PROCESS OBSERVED

An observer of the process leading up to and including the event of The Shift will most probably see the following:

- A gradual decrease in Earth's magnetic fields to near zero as a function of a reduced rate of rotation (longer days and nights), followed by a sustained electromagnetic null-state for approximately three days (72 hours).

- This time will be followed with the reinitialization and gradual increase of core-mantle-crust rotation, accompanied by the increasingly stable magnetics of a polar-reversed field. Earth's base resonant frequency will stabilize at a new resonance point determined by the Fibonacci scale.

- Unusual activity within other bodies of the solar system such as sun spots and solar prominence of record magnitude. Please note *Science News*, April 8, 1989, "Fantastic Fortnight of Active Region 5395," and the record solar impulses of January and March 1997.

- The radical shift/breakdown of energetic systems of "balance.
  > Geological—Record setting seismic activity accompanied by new types of faults, fault zones and earthquakes is demanding a rethinking of planetary models.
  > Meteorological—Unusual weather patterns, winds, ocean currents accompanied by extremes of record rainfalls and flooding, as witnessed in the Midwest during the fall of 1993, for example.
  > Social unrest—The appearance of chaos evidenced as "pockets" of unrest expanding and overlapping into other systems such as political and economic (record numbers of deaths reported in the U.S. from handguns in December of 1993).

Quantum mechanics tells us that even in the apparent vacuum created by this process, void of magnetics and oscillatory motion, *there may be continued fluctuating activity within the system*. Ancient texts tell us that this time of apparent nonactivity is actually a time of tremendous realignment. Within the subtle energy fields, the "blueprints" of Earth realign themselves to accommodate the new patterns of energy/information, within which it is being asked to exist. In Biblical terms, these three days in our future are equivalent to the process modeled to us during the three days before the Resurrection and following the "burial" of the Universal Christ.

THE PROCESS EXPERIENCED

The *experience* of an individual leading up to, and including, The Shift would reflect:

- A drastic change in the manner with which individuals, corporations, governments and nations perceive/relate to themselves, others, their lives, jobs, careers and family. A general sense of loss of control or lack of value for tasks completed. These changes may be attributed to the destabilization of the grids of emotion containing reference patterns for "old" paradigm values: home, family, career and life purpose. These patterns lose their meaning without the blueprints of the structures that they have been based upon. As these grids fall away, there is a period of readjustment accompanied by feelings of "unconnectedness," loss, and depression.

- The radical shift/breakdown of the systems forming the infrastructure of technology/society (social, political, economic, military, agricultural and industrial). As with the previous indicators, old patterns lose their meaning without the blueprints of the structures that they have been based upon. The fall of the Berlin Wall, the ending of the "Cold War" and the dismantling of dictator states are examples.

- A perceived increase in the passing of time. Events may feel as if they are "speeding up" due to the body's attempt to match the increasingly rapid pulsations of Earth's fundamental vibration.

- Intermittent loss of memory concerning "trivial" information. Birthdates, anniversary dates, bank-ATM codes may take on less of a sense of importance. Occasional episodes of a feeling of "unreality" while carrying out day-to-day activities. Though these kinds of information are perceived as "significant," their importance pales in light of the massive change that the body/spirit/mind complex is being asked to accommodate.

- Odd sleep patterns with periods of "null" sleep. Patterns of "black hole" sleep characterized by waking up physically exhausted after an adequate amount of sleep, feeling fatigued and tense, occasionally with sore muscles, followed by consecutive nights of vivid, prolific and meaningful dreams.

- Individuals experiencing tremendously difficult challenges as they draw into their lives their worst fears to be balanced/healed. The fears will be unique to each as a result of individualized experience. For some it is loss of finances, loss of spouses/mates or loss of health, and so on.

- During the event of The Shift itself, many individuals will experience the 72 hours in an unconscious and relaxed state akin to the Tibetan bardo state—essentially a dream experience. Those who have remembered the gift of Compassion may allow for the tremendous change while functioning within these new parameters. They will be present to facilitate the processes in others.

In and of itself, the experience of any one event does not signal The Shift. Several of these events are occurring at present, however, and may serve as indicators to the timing of the event itself. "When will The Shift occur?" is a question that is asked often. To the best of my knowledge, the precise timing of the event of The Shift is not known, even to the *"powers that be."* The intensity of the indicators point to an event very close at hand, *certainly* within this lifetime, possibly before our year of 2012. At the same time, from my perspective, those who live life waiting for The Shift will miss life! Our day-to-day lives in the presence of one another provide the very parameters that are teaching, training and preparing us to accommodate the tremendous change offered as The Shift of the Ages.

Though the "Shift to Zero Point" is a rare event by any standards, please remember that it is part of a very natural process; good, healthy and vital to *conscious* human evolution. Without exception, The Shift is the very reason that you have chosen to experience Earth at this particular time in history—it is the reason that you are here!

The last time that Earth underwent a complete 180 degree magnetic reversal was approximately 11,000 to 13,000 years ago. Evidence that this has happened in the distant past is seen as discontinuities in the geologic record, fossilized remains in the polar regions of North America and Siberia, as well as, accounts in the ancient texts. This time correlates closely to Plato's account of the breakup and submergence of the final portion of Atlantis in 9500 B.C. A physicist may call this period of time an electromagnetic null zone or a time of Zero Point, an idealized state of consciousness that is the goal of many meditative practices. Following this time of quiet, a new set of parameters begin to "drive" the processes of conscious life on Earth once again. Earth begins a new cycle of awareness; a New Age.

Following The Shift We May Expect

- The body of Earth, as well as the inner and outer cores, resume motion in the opposite direction relative to preshift conditions. The physical poles of the planet have reversed. What was once north becomes south and what was south becomes north. The reversal is a reference to the orientation of the magnetic fields only and not the actual sphere of Earth tipping 180 degrees.

- The sun will rise and set from opposite directions relative to its present cycle.

- Earth to settle into resonance with a new "tone." All matter, including biological life, has the opportunity to come into resonance with a new body of information, matching the new vibration or energy-information-light. Matter that is not capable of this match, either directly or through access to one of the harmonics, will disassociate into its Earth resonant, elemental constituents. This will be viewed as the great healing of the Earth. Compounds that are not naturally occurring (toxins, pesticides, additives and artificial created nuclear materials), having been forced together under artificial temperatures and pressures, may choose to disassociate.

- Human awareness to be experienced consciously, on multiple levels, simultaneously. This is a new state of consciousness with no name as yet. Rather than compartmentalizing experience into a predetermined brain state, such as beta, relaxed alpha or sleeping theta, delta or "K" states, all of these states are in simultaneous union providing an awareness of the totality of being. This is the beginning of the time referred to today as the "New Age" or by some as a fourth dimensional experience.

The reinitialization of Earth's rotation signals the beginning of a new interface, the phasing into a fourth dimensional experience. Considered to be a transition or "holding zone" for consciousness to assimilate the events that have just transpired, this transition correlates well with the references from *The Mayan Factor,* by Jose Arguelles, to Earth presently experiencing the final period of the great cycle from 1992 to 2012 A.D.[25] *During this period, Earth continues to express as in 3-space while experiencing resonance to 4-space.* It is at this point that Earth actually begins the *Ascension* process identified within the Mayan and Biblical texts; Earth migrating, in its entirety, into a more stable fourth dimensional experience lasting for approximately 1,000 years.

> *...This is the first Resurrection. Blessed and holy is he who shares in the first Resurrection. Over such the second death has no power, but they shall be priests of God and Christ, and they shall reign with him a thousand years.*
>
> THE HOLY BIBLE, REVELATION 20:5

Resurrection: A Shift in Dimensionality

A subtle, yet profound relationship exists between the energetic functions of the human Earth-grid-matrix systems. While this relationship may not be fully acknowledged by the traditional sciences as yet, it is suspected by those allowing themselves to move beyond the boundaries of their chosen discipline. The relationship must be viewed from a perspective beyond the constraints of flesh, fluids and membranes of man, and the mineral, rock and ocean perspective of Earth. Once again, the viewpoint must be from that of energy, the fundamental thread binding together the systems of man, Earth and all of creation.

With respect to the human-Earth relationship, each cell within the physical body of each human living upon Earth at present is attempting to align itself to a higher range of information, a zone of frequency operating at a greater pitch than that which has been experienced within the last 2,000 years of human history. This zone of high-frequency information represents an evolved awareness that each individual is striving to attain. It is this state of awareness that the Universal Christ attained during his 33 years on Earth. Through his living example of the use of choices made during his lifetime, and the use of free will to carry out those choices, Jesus of Nazareth was able to achieve resonance to a higher, more evolved body of consciousness that he referred to as "Father."

In effect, there are two shifts that are occurring, simultaneously: one of magnetics and one of frequency. Either shift may occur as a stand-alone event. As we see in the geologic record, the magnetic shifts have happened relatively frequently, to our knowledge without a dimensional shift. It is the increase in fundamental frequency that is driving The Shift in dimensionality. The timing of the magnetic shift to coincide with the dimensional translation is a "gift" that allows us direct access to ourselves to balance (heal) and accomplish the dimensional shift successfully.

The idea of vibrating each cell of the mind-spirit-body complex into a higher form of expression is not new and transcends the ideas of religion. The process itself has been modeled to us several times throughout history and has been recorded with varying degrees of accuracy. The three-day, 72-hour pyramid initiation within the Great Pyramid is a very good example, as well as the three-day, 72-hour process modeled by the Universal Christ during his time of Crucifixion, burial and subsequent Resurrection.

> *The Son of man is to be delivered into the hands of men, and they will kill him, and he will be raised on the third day.*
>
> *THE HOLY BIBLE*, MATTHEW 17:22

Each of these was a demonstration of a *conscious migration to another state-space*; another expression of being. The Biblical term for this process is that of Resurrection. Resurrection results from the conscious vibration of all aspects of the mind-body-spirit complex moving into a higher expression of the same body; a new expression of life.

Resurrection is life!

Resurrection is one path toward achieving a higher degree of vibratory expression, *through our bodies*, as each cell of our bodies is brought into perfect resonant harmony with a higher expression of information. Resurrection is not about our soul essence of life force leaving our body; that is death. This is why our body has always been, and is now especially important; our body *is* the temple that makes Resurrection possible. It is through our body that our soul may express the wisdom attained as our life experience. Resurrection is the conscious access to the next relative zone of experience ascending from the present. From a third dimensional experience, Resurrection may begin as resonance to 4-space–fourth dimensional information. Vibrating at such a high pitch, third dimensional matter no longer has the same meaning, or function, from the same point of reference as pre-Resurrection. The body is cleared of inharmonic patterns of energy that result in disease, aging and deterioration.

To learn through disease in the body is a choice!

All patterns of life energy within the physical body are at rest—balanced, as each aspect of belief, emotion, thought and feeling shift into harmony with the new reference point. The energy of reference is that of love and Compassion—the purest frequency of expression that may be generated and sustained within each cell of the body. The frequencies of love are not bound by dimensionality or time. It is a new expression of the energy of love that is the object of the Resurrection. This was the gift of Christ Jesus, the living demonstration of the tool, the life experience, as the bridge from the third dimensional experience of low frequency-high magnetics to the optimized experience of high-frequency and low magnetics.

It is Resurrection, and the subsequent Ascension, that is the goal of our life experience. We attract into life the experiences that provide us the wisdom to allow ourselves to "tune" to Earth's shift. As resurrected beings, we are in resonance (connected) to the energy (wisdom) of fourth dimensional grids, while still expressing within the context of a third dimensional world. We express as a very "high" being and may not be physically touched by others that have not progressed through a similar experience.

Activity	Frequency	Magnetics	Process Goal
Pyramid Initiation	Increase	Decrease	Immortality through conscious vibrational shift (Resurrection)
Christ Crucifixion	Increase	Decrease	Resurrection/Ascension The demonstration that life may be eternal as a result of choice
Earth Shift	Increase	Decrease	Dimensional Shift/Healing (Resurrection) The reestablishment of balance across all systems of energy

Table 4 *A comparison of current, as well as historic models, for Resurrection.*

In the case of many E.T. reports in recent years, those coming into contact with such a high-frequency being are subjected to patterns of energy that their internal circuitry may not be prepared for. These are the immortals of Biblical and ancient Egyptian references, those who had progressed through their spiritual evolution to the point that they were able to transform themselves into the pure energy of the next state space.

Earth is undergoing a process identical to that modeled by very evolved beings in Earth's history: the Resurrection or vibration of each aspect of the self into a higher expression of experience. It is no longer necessary to go into the sacred chambers of ancient structures designed to provide the appropriate environment for this Shift; *we are living The Shift within our offices, schools, living rooms as day-to-day life.*

Through our life experience, we are already preparing for, and learning, the process of Resurrection! Each activity that we do, every relationship that becomes a part of our life, regardless of how trivial it may seem at the time, is providing a tool for us to assimilate, generate and sustain higher and greater frequencies of energy-information-light. Each experience provides a thought or feeling. That feeling may be considered as a packet of information-energy, seeking balance somewhere within the framework of what it knows; the matrix of our mind-body-spirit complex. Each time we have an experience, there are thoughts and feelings associated with that experience. They are part of us and can never be "forgotten" or erased. As we interpret those experiences through the lens of our life-tools, we make a determination as to how that experience will be stored. To the degree that we have chosen to place the energy of each experience in its place of balance, to that degree do we become a being of balance.

Awakening then, is a term that accurately describes the cumulative series of processes, spanning hundreds of thousands of years, affecting all life upon Earth. Awakening is the beginning of The Shift; the collective initiation into the time of a new wisdom—the wisdom of the feeling heart.

THE RAPTURE: WAVES OF ASCENSION

Prior to the time of the Collective Initiation, an opportunity will be presented for some individuals to experience the Resurrection process early—possibly *days* before the actual Earth Shift. This event, prophesied to occur at the close of the cycle, is known as the time of the Rapture. Intended as a time of celebration signaling the completion of this cycle of awareness, for many the process conjures up images of fear, as they witness the familiar world in dissolution. The Rapture is the beginning of a new experience for those who have loved Earth, and the people of Earth, to the degree that they have offered lifetimes of themselves to the process of this conscious cycle. The events of this process mark the completion of that period and the beginning of the *Resurrection-Ascension* process into higher dimensional experiences, for Earth as well as the individuals.

The opportunity to experience the Rapture will not be perceived in the same manner, at the same time, by everyone, just as not all souls incarnated upon Earth at this time are progressing at the same soul-development rate. Due in large part to the operation of Choice and Free Will throughout this conscious cycle, each individual has the opportunity to experience and evolve at a pace that is suitable for that individual. As the cycle draws to a close, those that have accelerated will have the opportunity to complete their process "early," that is, prior to the Earth Shift. *There is no intended judgment of right or wrong inherent in the ability to recognize the Rapture; it is purely the completion of one aspect of experience.*

Individuals that have incarnated relatively recently into the memory-pool of human awareness, "seeds," have done so in an effort to infuse into the conscious matrix a renewed understanding and sense of vitality regarding the purpose, timing and function of humankind within this cycle. The Biblical texts tell us that there were originally 12,000 seeds that incarnated into each of the original 12 tribes of Israel, offering different families, different customs, different vocations and carrying those attributes to a variety of geographic locations. One hundred and forty four thousand individual seeds, each anchoring a unique expression of a common body of information. In contrast to the single perfected seed of a "Christ," the many seeds optimize the chances of successful "germination" of a Christed vibration within the matrix. Because the seeds have not experienced within the Earth incarnation loop for lifetime upon lifetime, they are less likely to incur heavy karmic debts and the tremendous feelings of loss and separation felt by those of the original seeding. The intent was to make use of the holographic nature of the conscious matrix and once again anchor the memories of purpose, clearly and purposefully, at a time close to the completion of the cycle. It is to these individuals, the seeds, and any others that have allowed for the memory of their truest nature, that the term of The Rapture becomes meaningful.

The Rapture may be defined as both a time, and a process, within which specific individuals have the opportunity, and may *choose* to respond, to a series of conditions that allow for their immediate and simultaneous Resurrection and Ascension.

Individuals who may experience the Rapture are preconditioned to recognize the codes, though they may not be aware of them at this time. They will "feel the call" as a series of pulsed vibrations-tones that will be recognized on a level beyond that of conscious knowing. Through a harmonic resonance achieved as the result of the life experience, their bodies will be "tuned" to come into harmony with the tones, the Biblical "trumpets" of Revelation. With this resonance as a key reference vibration, specific fields tuned to the resonance will have the tendency to vibrate much quicker. The tones will essentially allow specific fields surrounding the body, those of the human *Mer-Ka-Ba* (Chapter 2) to move into speeds and key ratios of vibration that will accelerate the experience of the dimensional shift for those individuals. If they choose, *if they know or learn this tone*, the familiar motion of the *Mer-Ka-Ba* fields will allow them to access the vibrations of the fourth dimensional state space preceding the collective Earth experience.

> *No one could learn that song except the 144,000 who had been redeemed from the Earth.*
> THE HOLY BIBLE, REVELATION 14:3

At this time, the close of the cycle, some individuals will have fulfilled their commitment to usher Earth into a New Age and will *choose* to no longer remain through The Shift. For these individuals, the Rapture will signal an opportunity to "return home" to other dimensional corridors and experiences. Others, even in close proximity, may not be capable of hearing the tones; *their promise is to complete the cycle with the Earth*, seeing it through from the beginning to the close. Again, the ability, or inability, to perceive and respond to the tones of the Rapture must not be judged through feelings of self worth: "Am I good enough for this experience?" For some individuals, an inherent portion of the life path will include the Rapture. Others will remain with their energy firmly anchored within the Earth/Shift experience. The processes and emotion generated as the Rapture may represent the greatest final challenge to consciousness, providing an opportunity to experience without judgment. Each individual, in journeying with "family" through the experience, will complete this cycle in much the same way in which he or she has come into this world: as a group, and also alone.

Each will have the opportunity to face his or her fears as they witness the seeds disappear as instantaneous Resurrection and Ascension. Each will have prepared, some for many lifetimes, for this very moment. Each will know on a deep level that will be instantly accessible to them in that moment, as to the appropriateness of allowing the new tones of the song to move their fields into motion. It will happen quickly and there will be no indecision.

At this point I would like to relate an experience that I had during a journey to Egypt in 1987. I was with one other individual in the city of Luxor during an afternoon walk toward the marketplace. From no apparent direction an elderly Egyptian man in traditional dress, thin with dark skin and very clear eyes, walked up to me, looked directly into my eyes and, without the formality of greeting, spoke in very good English the following words:

> *Some day soon, you will be asked to make a decision. You will have no time to think about it. Are you ready?*

As this gentleman walked away, I was left reeling in the street from the intensity of the experience. This information was relevant to me at the time and to this discussion of the Rapture. At this point I knew that there would be multiple Raptures preceding multiple resurrections leading up to and including The Shift, each offering a specific tone directed to those tuned to its vibration. The Rapture(s), though designed to provide a gentle boost to those seeds who have completed their tasks within this great cycle of experience, are available to all who may come to remember the tone. They may consciously respond to the familiar vibration that brought them into this experience long ago. The feeling of the tones will evoke great emotions of love and will be focused upon the seat of the Time-Space-Light Vehicle known as *Mer-Ka-Ba*—anchored within the heart (Chapter 2). The Biblical texts refer to multiple Resurrections—implied through the reference to a *first*—over a period of time as Earth progresses up to and through The Shift.

> *And now, the Resurrection of all the prophets, and all those that have believed in their words, or all those that have kept the commandments of God, shall come forth in the* first Resurrection; *therefore, they are the* first Resurrection.
>
> THE BOOK OF MORMON[26] MOSIAH 15:22

> *...They came to life and reigned with Christ a thousand years. The rest of the dead did not come to life until the thousand years were ended. This is the* first Resurrection. THE HOLY BIBLE, REVELATION 20:4

Those having chosen this path, that of the Rapture, may just now be awakening to the possibility of who they are. With the awakening, those same individuals may begin to feel a sense of urgency in their life-path. Even as seeds, many have become "lost" in the density of the Earth experience and are just developing a new sense of identity. As the memory returns, the constructs of their lives may appear to make less sense; careers, relationships and even commitments made prior to the wake up, may not feel harmonious with the codes being activated from within; the patterns simply will no longer "fit." For those individuals, their path will lead them through the doorway of experience that brings them to the threshold of resolving those feelings, facing the fears and coming to terms with the emotion of what has come to pass. Even now they are preparing, and being prepared for, the Rapture.

SUMMARY

- Events of unprecedented change are occurring upon the Earth now, and are seen as dramatic shifts in the historic behavior of social, political, economic, military, geological, meteorological and atmospheric patterns of energy. All are shifting in an effort to match new patterns of cyclic energy-information-light that Earth is approaching.

- The understanding of these shifting patterns is not religious or metaphysical in nature, although entire religions and belief systems have developed around the "truths" and significance of these processes and this time in history.

- Ancient systems of time keeping (Tibetan, Chinese, Biblical, Mayan and those of indigenous peoples) point to this time in history as being significant, both in terms of human and planetary patterns.

- Two fundamental indicators that delineate this time in history are digitally measurable parameters of Earth magnetics and base resonant frequency.

- As Earth moves through the change, known as "The Shift," the chemical composition of the planet will remain the same, while the structural (geometric) expression of the energy reflects the change; we will remain carbon-based beings on a silica-based world.

- Modern physics describes the point at which the amount of vibrational energy associated with matter under certain conditions to be that of "Zero Point." Earth is experiencing the early stages of events that will provide an experience of Zero Point, "The Shift," allowing the breakdown of thought constructs that are inharmonic with the "blueprint" of expression.

- Resurrection is a nonreligious term referring to the conscious vibration, a phase transition, of each cell of the body-mind-spirit complex into a new expression of itself. To accomplish this process, fields of magnetics decrease to zero, while the fundamental vibration increases to a threshold resonance point. Earth is living this process at present (as of this writing).

- Lower fields of magnetics provide the opportunity to manifest rapidly, as the interference patterns of the magnetic fields subside.

- The fundamental relationship between frequency (sound) and geometry (form) is two-way. Form yields sound and sound yields form.

- The process of The Shift may well occur over a three-day period, following the models presented through initiations and accounts of previous Shifts detailed in the ancient texts.

- Our life experience is the primary tool of remembering to match Earth resonance, and move, gracefully, through "The Shift of the Ages."

CHAPTER TWO

The Language of Creation

MATTER, ENERGY AND LIGHT

Know ye, O man, that all that exists, is
only an aspect of greater things yet to come.

Matter is fluid and flows like a stream,
constantly changing from one thing to
another.

The key to worlds within thee,
is found only within,
for man is the gateway of mystery,
and the key is that One is within One.

ADAPTED FROM *THE EMERALD TABLETS OF THOTH*

Human history is laced with prophecy, warning and predictions of an event that will forever alter the lives of every man, woman and child existing upon the planet at the time of the event. Within each culture there are organizations, societies, sects and orders that claim to have preserved a set of traditions or belief systems to be used in preparation for the "event." In Western traditions, these societies have become religions. Socially oriented organizations frequently offer the ritual of ancient traditions without explanation, or in some instances, without the knowledge of why the rituals are performed. Incomplete teachings are offered as a means of preparing for the event that signals the "end times." The belief is that by following these rituals daily, going through the motions, the individual has somehow prepared for this turning point in human consciousness. Though distorted to some degree in their present form, the mystery schools were created in an effort to preserve the integrity of the ancient knowledge for a time when it would be needed, just prior to the close of the cycle. An integral portion of the beliefs inherent in any of the mystery schools may be found in the knowledge that Earth and the human form are intimately entwined. Specifically, the fields of energy-information-light surrounding each cell of the human form, and all 3-space matter in general, are possible only within a range of parameters offered through Earth at present.

Within the context of The Shift of the Ages, any meaningful understanding of consciousness and our present paradigm must take into consideration the physical processes of Earth. Both the human and planetary matrices express as cells, bone, mineral and ocean. All are dense manifestations of a more subtle and fundamental expression of creation, pure energy expressed as vibration (oscillation or frequency). You may think of these vibrations as *pulsed waves* created by discrete bursts of energy, fundamental frequencies often referred to as light. Modern researchers recognize these bursts of energy as *quanta* (brief, rapid pulses of light) and study these phenomena as the science of quantum physics.

The term *light* is used broadly throughout history representing any number of concepts. For example, light is often a generalized term referring to "good"; something that is "in the light" or "from the light" is desirable. The Biblical references to light, beginning in the Book of Genesis, relate the workings of the creative energy referenced as God:

> *And God said, let there be light, and there was light. And God saw the light, that it was good, and God divided the light from the darkness. And God called the light Day, and the darkness he called Night.*
>
> *THE HOLY BIBLE,* GENESIS 1:3–5

In the New Testament, references to light through the teachings of the Universal Reference Being, Christ Jesus, refer to light as a *force within the individual* expressed as deeds performed throughout a lifetime:

> *You are the light of the world. Let your light shine before men, that they may see your good works.* *THE HOLY BIBLE,* MATTHEW 5:14

Additional references to the light from within may be seen in Tibetan texts:

> *Thine own consciousness, shining, void, and inseparable from the great body of radiance, hath no birth, nor death, and is the immutable boundless* light.
>
> *THE TIBETAN BOOK OF THE DEAD,* PADMASAMBHAVA

As well as *The Emerald Tablets of Thoth:*

> *Thy* light, *O man, is the great* light *shining through the shadow of flesh, free must thou rise from the darkness, before thou are one with the* light.
>
> *THE EMERALD TABLETS OF THOTH,* TABLET 9

> *Darkness and* light *are both of one nature, different only in seeming, for each arose from the source of all. Darkness is disorder,* light *is order; Darkness transmuted is* light *of the* light.
>
> *THE EMERALD TABLETS OF THOTH,* TABLET 15

More recently, light has become a term referring to radiant wavelengths of electromagnetic information that are perceptible to the human eye. This range is typically depicted as the colors of the rainbow with a hierarchical order of ascending frequencies measured in very small wave units of Angstroms (Å).* Beginning just above the infrared range, red is perceived between 7,700 Å to 6,200 Å, orange at 6,200 Å to 5,920 Å, yellow at 5,920 Å to 5,780 Å, and so on (Fig. 17). For the purposes of this text, the terms of energy, light and information may be used synonymously.

Within Biblical references, as well as other ancient texts, the term light refers to the *full spectrum of electromagnetic information,* including and not limited to, visible light; *all*

*One Angstrom unit is equal to .00000001 centimeters.

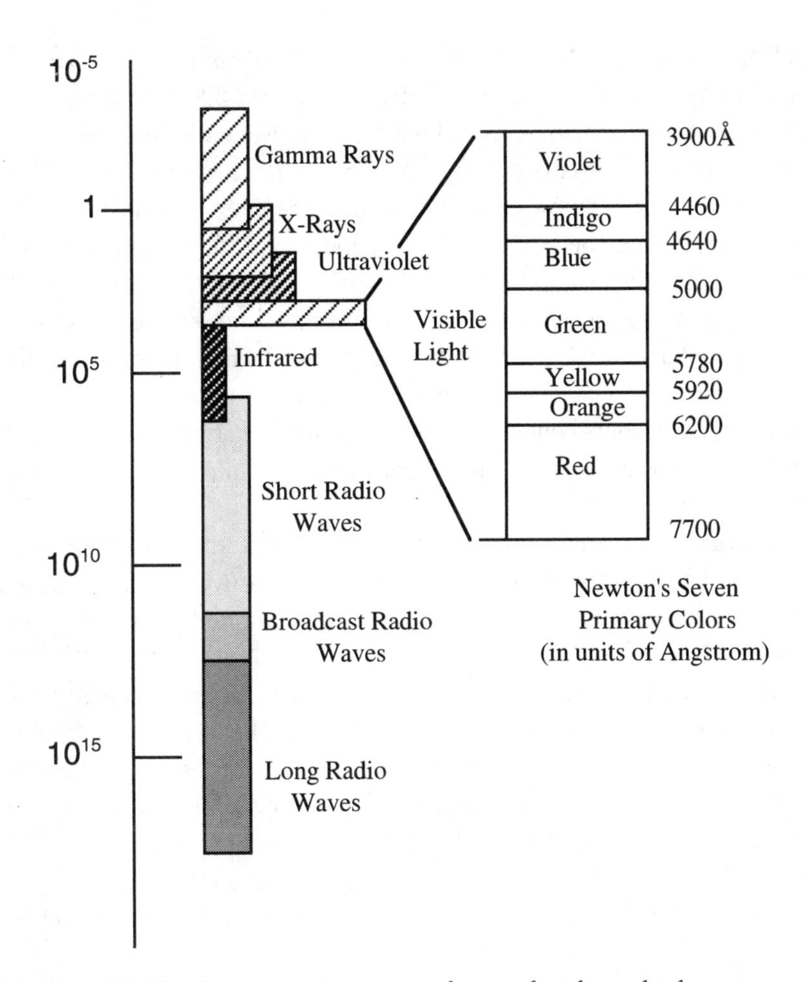

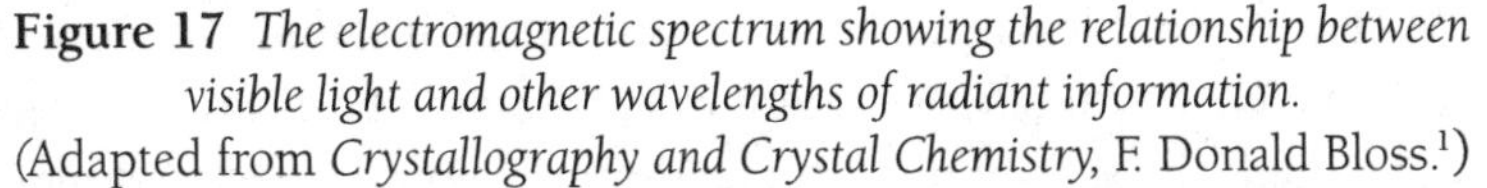
Figure 17 *The electromagnetic spectrum showing the relationship between visible light and other wavelengths of radiant information.*
(Adapted from *Crystallography and Crystal Chemistry,* F. Donald Bloss.[1])

wavelengths of radiant energy permeating all of creation throughout all universes. Light *is* energy. Light *is* information. Light *is* the fabric of creation. Under given conditions, light will express itself repeatedly and consistently in response to those conditions. The expression may be seen as a speeding up or slowing down of its essence; an appropriate crystallization in response to a given condition. Within this context then, the "daylight" of the early moments of Biblical creation may be viewed as the congealing of many wavelengths of radiant information, of which visible light is one portion.

Building Blocks of Light

Light may be defined in a number of ways, each accurate within the context and intent of its use. Visually, we know when we see light, when we do not, and varying degrees of in-between light that are detected through the faculty of sight. Technically, we are "seeing" within a very narrow band of the many possibilities of vibratory information, very short and rapid pulses of energy typically between 7,700 and 3,900 Å in length. For some individuals, the entire experience of life is based upon visual events occurring within the constraints of this limited band of vibration 3,800 Angstrom units wide. What about events occurring beyond this range—above 3,900 and below 7,700 cycles per second? What about events that are occurring within regions of vibration that are so rapid, they escape detection by sophisticated instruments of measure?

A clue to this may be found within the memory of large super computers simulating and modeling the creation of the known universe. Using Cray™ super computers and extremely sophisticated mathematics, scientists have developed models of creation from the assumed instant of the theoretical "Big Bang," when the present universe originated and began to expand very rapidly. Allowing for error in the absolute values of the numbers used, researchers believe that they have good relative values for the amount of matter present in the universe at the instant matter began to move away from its existence as a single point.

Simulating this event within the computer, a mysterious phenomenon develops quickly. Shortly after the beginning a full 90% of the universe becomes unaccounted for leaving only 10% of the mass of the universe accounted for within the computer model and in observations.[2]

At the same time researchers in the life sciences ask us to consider that for any given individual, only a fraction of the brain is used. The function of the remainder is unaccounted for and appears as dormant. There are theories as to multiple, redundant biological circuits and a yet-to-be realized state of evolutionary maturity when the brain is utilized more completely. The numeric estimates remain. Only 10% of the human brain is utilized and only 10% of the mass of the universe can be accounted for (Fig. 18). Is it by chance that these percentages correlate so closely? Possibly not.

What are the computer models and biologists showing, or failing to show us? Neither the model, nor the life scientists are taking into account one of the *most fundamental* and possibly *least understood* dynamics of creation: the component of dimensionality.

The purpose of 90% of the human brain is unknown. 90% of the mass of the universe is unaccounted for.

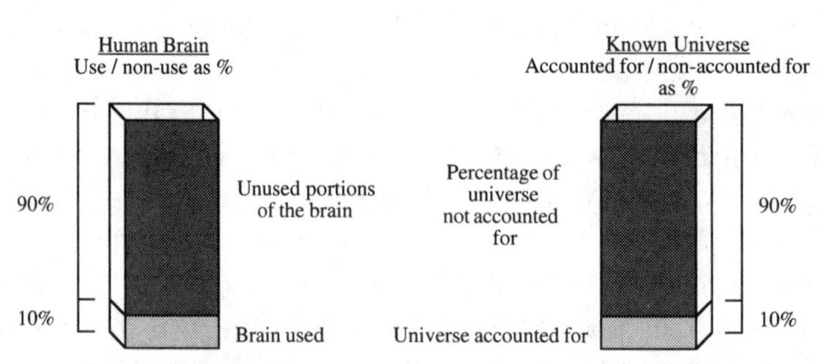

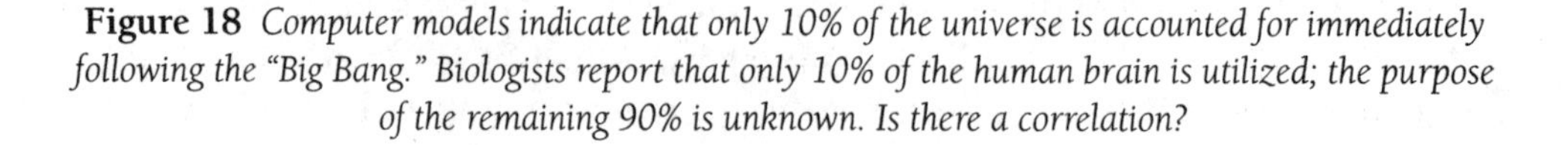
Figure 18 *Computer models indicate that only 10% of the universe is accounted for immediately following the "Big Bang." Biologists report that only 10% of the human brain is utilized; the purpose of the remaining 90% is unknown. Is there a correlation?*

Neither of these observations takes into account the parameters of dimensionality; events occurring at such high vibratory rates that they appear beyond our range of physical perception. The ancient texts remind us that for any individual, 100% of the human brain is in use, though perhaps not 100% consciously.

Shortly after the moment of creation, the universe was expanding so rapidly that its vibration could no longer express within the parameters of the three dimensional experience. In this possibility, 90% of the universe literally vibrated itself into higher dimensional state spaces! We are still tuned to the events occurring within those spaces through 90% of the resonator that we developed for just that experience; our brain. The 10% of the brain that we acknowledge as being used is accessing only the 10% of creation that we call the third dimension.

Delta	***Theta***	***Alpha***	***Beta***
.5–4 Hz	4–8 Hz	8–12 Hz	12–23 Hz
Deeply relaxed	Relaxed/ aware	Waking state	High alertness

Table 5 *Frequencies of Awareness.*
(Courtesy of Tom Kenyon and *Acoustic Brain Research.*[3])

Consider the comparison of human states of conscious and the base resonant frequency of Earth. Historically, Earth has resonated at approximately 8 Hz, or cycles per second. If Earth frequencies were considered from the perspective of human brain function, the vibrations of 8Hz would be considered as that of the alpha state—deep relaxation or sleep. The ancient texts, as well as the teachings of the indigenous people, tell us that is precisely where many humans have experienced as a consciousness, in a relatively unaware and unconscious state of sleep. Chapter 2 indicates that the target value of Earth's frequency may be 13 Hz, the next value found in the Fibonacci sequence from the historical 8 Hz. To the human brain this value is that of beta consciousness, the waking state approaching hyperwakefulness and high alertness. As we collectively move toward the vibration of 13 Hz, Earth, and all life upon Earth, is just beginning to wake up.

Traditionally the goal of meditation in general, and specifically any zero point meditation, has been to create the parameters necessary to access events occurring beyond those recognized by the active 10% of the brain. At present, however, as the close of the cycle draws near, a new awareness is coming into focus — an awareness that has no name as yet. This new state of consciousness results from simultaneously being awake within the remaining 90% of the brain, as well as the 10% historically used. Researchers are seeing this state in full-conscious channels working through a feedback device known as the Mind Mirror™. Typically, individuals active in delta and theta states appear to be asleep. The conscious channels, however, are active in the high-frequencies of beta, high beta, K-complex and above, as well as the lower frequencies. Without experiencing any one brain state discretely, the channels are accessing all brain states simultaneously as a new, and yet unnamed state of completeness: Zero Point Consciousness.

To develop a conceptual understanding of thought-matter moving through these vibrational membranes, it is useful to build a common vocabulary from which to refer to throughout this discussion—including the terms of *grid and matrix, wave and resonance.*

THE GRID: INFORMATION IN TWO DIMENSIONS

The term "grid" is a reference to a two dimensional framework, providing a preferred pathway for energy-information-light to travel from point A to point B. The energy of consideration may be as subtle as the microhertzian pulsations of human thought generated at hundreds of thousands of cycles per second, or as dense as Earth-resonant tones pulsating as low as 7.8 cycles per second. The energy, traveling as waves, is propagated throughout various aspects of creation along this lattice of high conductivity, transgressing the boundaries of star systems and dimensionality.

The grid may be thought of as the substance that exists in the nothingness, the thread of fundamental intelligence that is the underlying fabric of creation.

Conceptually, a grid may be considered as an etheric network of guidelines, a meshed framework along which pulses of energy are directed. These ordered patterns of energy are typically formed of a single, uniformly shaped pattern, repeating itself over and over as equally spaced expressions of the identical geometry in any two-dimensional direction. The form may be as simple as that of the cube or rectangle (Fig. 19) or as complex as an equal-sided polygon matching perfectly with the polygons surrounding it. The smallest single unit of space, enclosed within all directions by an equal-sided form of the same expression, is termed a "grid cell." Multiple grid cells, adjacent to one another and extending in all directions for some distance, finite or infinite, comprise the grid. Energy may travel, or be stored, along the boundaries of the framework or within groupings of the grid cells themselves.

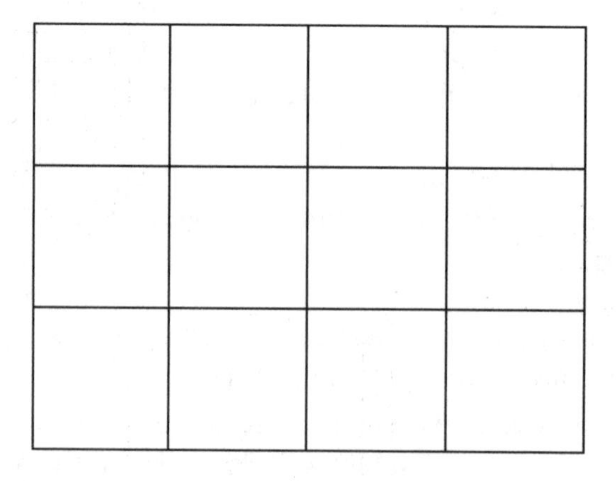

Figure 19 *Schematic representation of a finite, two dimensional grid based upon the Cube. One Cube represents a single grid cell.*

Conduits in Three Dimensions: The Matrix

The term "matrix" describes a collection of grids, appearing as stacked upon one another. The grids are hierarchical in nature, providing a structure for the gradual transition of energy-information from one zone of parameters to another (Fig. 20). The size, dimensions or lateral extent of the grid cell and overall grid *may* vary. The size and dimensions of the matrix itself *will* vary. Most matrices will be recognized as a subset of a larger grid-matrix system, and that one in turn will be a portion of an even larger parent system. For the purposes of this material, the grid-matrix relationship is a conceptual model built to understand the framework of creation and all life contained within. The matrices of creation are essentially *holographic* in nature; each cell being whole and complete unto itself, as well as part of an even larger whole. Within each single cell lives all of the information for the entire pattern to repeat itself again, and within each cell of that pattern the repetition continues (Fig. 21).

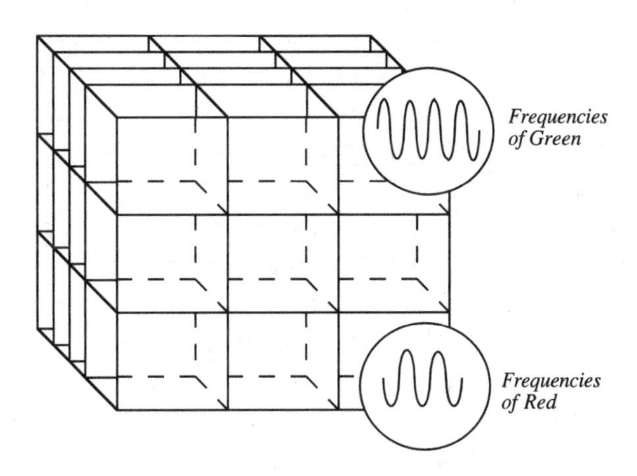

Figure 20 *Sample matrix composed of grids based upon the Cube. Each grid measures three cells by four cells. This is a uniform matrix of equal-spaced and equal-dimensioned cells providing a simple pathway for information to be transferred in either vertical direction (up or down) or horizontally within an individual grid.*

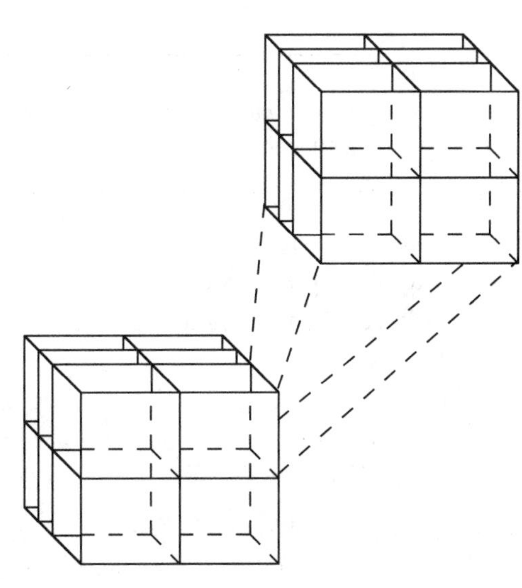

Figure 21 *Sample holographic matrix illustrating the concept of patterns within patterns. Each grid cell is complete unto itself, while containing all of the information required to reproduce the pattern of the "whole."*

Each grid may support information of a form that varies slightly from the neighboring grid above or below. Grids may be viewed as ascending or descending, depending upon the orientation of the viewer. For example, in Figure 20 the bottommost grid may support a vibration or frequency associated with the visible light range known as red. Ascending, the next grid could support a range of energy associated with the color of green. The inset illustrates this through a close-up of the wave form within one cell of each grid. The cell from the red grid shows fewer complete wave forms per unit cell.

It is vibrating slower than the energy within the cell taken from the green grid. A matrix, then, may be two or more dissimilar grids with multiple submatrices forming the whole.

INFORMATION AS WAVES

Within the open literature, the energy of light is typically shown in two dimensions as depicted in the cross section at the top of Figure 22.

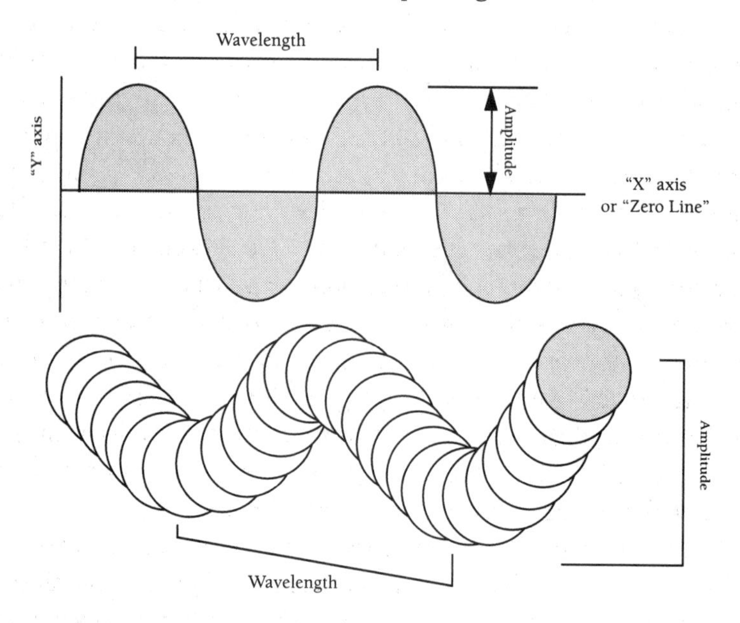

Figure 22 *Schematic representation of energy expressed as a wave. The lower illustration offers the concept of a many-dimensional wave, projecting as a sphere, simultaneously, in all directions.*

The illustration shows a single wave within a common frame of reference; the X/Y graph. The vertical bar represents the Y axis and will be used to indicate the distance above or below a fixed reference point to which the wave extends. Horizontally, the X axis measures the distance from the beginning to the end of the wave or the "wavelength." The height, or distance, from the zero line of reference to the top of the wave is termed the "amplitude" and may be thought of as representing the intensity of the wave. Frequency is a term often used in "New Age" phraseology and may take on different meanings within the context of the conversation. For the purpose of this material, frequency describes the number of complete waves to pass a given point per unit time.

For example, consider the intersection point of the X and Y axis as the point-of-reference, or origin point.

The number of complete waves passing through the origin point in one second is referred to as the frequency.

Ten waves passing the reference point in one second would be 10 waves per second, commonly referred to as 10 Hertz (Hz). Values of frequency may range from very slow vibrations of fractions of 1 Hz to extremely high-frequencies of hundreds of thousands of waves per that same second. The same energy moving at different speeds and expressing differently because of the speeds.

All matter whether dense physical or less dense, nonphysical, is essentially composed of the same fundamental building blocks of energy, *light*, expressed as hierarchical fields of interlocking vibration (frequency) and form (geometry). A portion of these fields is described in the science of physics as the electromagnetic spectrum and range from long waves vibrating very slowly at the bottom of the spectrum, to very short, rapid waves measured as fractions of a centimeter at the high end of the spectrum (Fig. 17). Generally, the electromagnetic spectrum today is not shown as vibrating much beyond values of 10,000,000,000,000,000,000,000 Hz (10^{22} cycles per second). During the late 19th and early 20th centuries, however, scientists such as John Keely routinely worked with vibrations ranging *20 to 30 octaves above* those recognized in the open literature.[4] Through working with these subtle vibrations Keely was able to develop technologies that were not well understood by his contemporaries, and are not even acknowledged today.

Within the context of our third dimensional experience, the energy of light exists as two separate, though related, waves. These two distinct components function together to provide the physical expression of more subtle aspects of the creation matrix: the *electrical field* and the *magnetic field.* When considered together, the two components provide a single field known in Western thought as electromagnetic energy expressed as a wave (Fig. 12). Without exception, all physical matter may be viewed as an expression of electromagnetic fields of energy, or "light."

The electrical portion of the wave may be considered as the carrier of information. It is upon the carriers that the codes of creation, literally crystallized light, are propagated from one expression of creation to the next. Each specific frequency of light, experiencing within a range of vibration and magnetics, produces a consistent, repeatable and meaningful pattern of crystalline codes. These are the specific codes that are picked up within the genetic structure of life, the instructions providing information to the cells allowing them to replicate consistently and repeatedly.

Though related, the magnetic subwave trails behind the electrical portion of the wave, never completely overlying the information portion of the signal. The magnetic aspect lags at 90 degrees relative to the informational aspect of the composite wave (Fig. 12). It is this lag that creates the buffered interference zone around the electrical information, stabilizing that information as well as preventing complete access to one field of information from another.

It is within this buffer that our interference patterns of fear and judgment reside. It is this buffer that is deteriorating as Earth's rotation slows, allowing us a rare opportunity of direct access to ourselves as information in the absence of old beliefs, rituals and conditioning.

DIMENSIONALITY: A SHIFT IN VIBRATION

Dimensionality may be viewed as an attempt within the logical mind to resolve the hierarchy of energy into discrete bands, or ranges of experience, so that they may be referenced separately. The reality is that there is no separation. There are no clear boundaries between one vibrational experience and another. There is only a gradation, a phasing of varying states of experience. These altered states of awareness are known as dreams. Though deceptively simple in appearance, this premise of "oneness" has proven difficult to resolve within the rational mind. The ancient texts however have offered this very concept, through many different languages, for thousands of years.

> *...If the mind makes no discriminations, the ten thousand things are as they are, of single essence. To come directly into harmony with this reality just simply say when doubt arises, "not two." In this "not two" nothing is separate, nothing is excluded. In this world of Suchness there is neither self nor other-than-self. To understand the mystery of this One-essence is to be released from all entanglements.*
>
> HSIN HSIN MING: *SENGSTAN'S VERSES ON THE FAITH MIND*[5]

> *When you make the two one, and when you make the inside like the outside and the outside like the inside, and the above like the below, and when you make the male and the female one in the same...then you will enter the kingdom of God.*
>
> *THE GOSPEL OF THOMAS*[6]

> *Look thee above, or look thee below, the same shall ye find, for all is but part of the Oneness that is at the Source of the Law. The consciousness below thee is part thine own, as we are a part of thine.*
>
> *THE EMERALD TABLETS OF THOTH,*[7] TABLET 11

Dimensionality may be considered from the perspective of one substantive energy, expressing itself differently in response to changing conditions; those within which frequency may increase or decrease.

Creation, viewed through the lens of pattern, is simply a collection of waves—some moving much faster than others, expressing as interlocking patterns of energy, *very slow crystalline waves*. Everything that we have ever physically touched, built, created or uncreated has ultimately been possible because dense light has been manipulated into a new expression; a recombination of energy crystallizing into one or some combination of patterns that *are* the light within those parameters. All that exists within creation, vegetable, plant and mineral, inclusive of all life and all that we build is what it is because of interference patterns of waves interacting with waves. We may produce waves on a gross scale and influence the world around us, as demonstrated at the turn of the century through Nikola Tesla's oscillation machines. We may simply think and feel with the sincere conviction in the belief of life force and generate high-frequency waves (love) from the internal oscillator of the human body.

> *Within the gross vibration of flesh is the fine vibration of the cosmic current, the life energy; and permeating both flesh and life energy is the most subtle vibration, that of consciousness.*
>
> *The "word" is life energy or cosmic vibratory force. The "mouth of God" is the medulla oblongata in the posterior part of the brain, tapering off into the spinal cord. This, the most vital part of the human body, is the divine entrance (mouth of God) for "word" or life energy by which man is sustained.*
>
> SCIENTIFIC HEALING AFFIRMATIONS,
> PARAMAHANSA YOGANANDA[8]

> *Man shall not live by bread alone, but by every 'word' that proceeds from the "mouth of God."* THE HOLY BIBLE, JOHN 4:4

The "word" of John is the life energy of Yogananda, proceeding from "the mouth of God," the base of the brain that the ancients referred to as "the seat of consciousness." All may be considered as different "textures" of vibration expressing as waves. All vibration is expressing as energy in motion—pulsating and oscillating as waves.

> *In science today we are witnessing a general shift away from the assumption that the fundamental nature of matter can be considered from the point of view of substance (particles, quanta) to the concept that the fundamental nature of the material world is knowable only through its underlying patterns of wave forms.*
>
> *SACRED GEOMETRY; PHILOSOPHY AND PRACTICE,*
> ROBERT LAWLOR[9]

In recent years, the Western concept of three dimensional matter as "solid" has gradually given way to the idea that the appearance of solid matter is an illusion. From this perspective, reality appears to be a series of rapid bursts of energy pulsing at specific rates to produce predictable and consistent wave forms. It is these patterns of information that our brain "sees" and averages together to produce the perception of a solid, continuous event. The indigenous peoples of the world have told us for thousands of years that our world is the world of illusion; that all in this world is the product of our thought and as such is not real. Additionally, ancient texts remind us that our world is actually the world of a dream, and at this precise time in Earth and human history, the dream is about to change.

Matter: Crystalline Light

The energy of creation is pulled down through the molds of the matrix, vibrating slower and slower, literally "crystallizing" into the familiar forms of the third dimensional experience. What are these pre-defined patterns that energy aligns with? How do they influence the formation of matter?

There is a fundamental relationship that exists between frequency (vibration) and geometry (form). The relationship may be viewed as functioning in two directions:

Frequency <==> Geometry

Frequency yields geometry and geometry yields frequency. Stated another way, form is a direct result of vibration and vibration is a direct function of form. *All matter exists because the energy of that matter is held in place as the substance of vibration.* This concept is not new. Ancient texts and mystical traditions have offered this premise for centuries. Western science is now remembering this law, proving to themselves one half of the concept: the consistent and predictable effect of vibration upon matter.

A simple demonstration that you may perform on your own will help to illustrate this sometimes nebulous relationship between sound and form. For the demonstration you will need:

1. A flat plate, such as a dinner plate or cookie tin
2. A granulated substance such as salt or sugar
3. Some method of reproducing sound
   (a transducer such as a stereo speaker)

Place the dinner plate directly over the cone of the speaker. You may have to lay the speaker on its back. Place enough of the granulated substance in the dish to completely cover the bottom with a thin layer. Through the speaker, play a series of prerecorded tones and observe what occurs to the powder upon the dish. An interesting phenomenon develops; a phenomenon that illustrates one of the most subtle, yet fundamental relationships throughout creation. The grains will rearrange themselves into patterns in the bottom of the dish, as if aligning to some unknown framework.

Specific tones, or combinations of tones,
will always produce specific patterns or combinations of patterns.

These will be predictable and repeatable. As you experiment with varying intensities of volume, bass, treble and single as opposed to multiple tones, you will soon discover that although the patterns may vary, cycles similar to those in Figure 23 will appear. They will originate as simple geometric patterns, evolving into more complex patterns, only to break down and begin a cycle again with less complex patterns.

Figure 23 *A solaris figure resulting from Cymatic frequencies. Specific frequencies, or combinations of frequency, produce consistent and predictable patterns in a given media. Higher frequencies tend to produce more complex patterns.* (Adapted from *Cymatics* by Dr. Hans Jenny.)

The study of these relationships is known as the science of "Cymatics." Much of the pioneering work in the field of cymatics has been through the efforts of Dr. Hans Jenny.[10] In the relatively small scale of the laboratory setting, Dr. Jenny has demonstrated that the interference patterns resulting from vibration offer a beautiful metaphor, describing the forces that drive creation. What appear to be chaotic systems of fluids within fluids, or particles upon plates, unorganized assemblages of matter, find order in response to vibration. Earlier within this text it was stated that "... all that exists within creation is what it is because of interference patterns of waves interacting with waves." It becomes apparent in viewing Cymatic relationships that the structure of creation, that which we see as our familiar world, results from energy responding to more fundamental guidelines, information aligning itself to some unseen pattern in response to subtle forces. The implications of this understanding are vast and an integral portion of the process of The Shift.

Creation exists, portrayed as energetic patterns of wave form, aligning along predefined guidelines of expression. As the tones of creation propagate, throughout the various environments of the universe, they produce predictable, repeatable expressions in response to those tones. We see those expressions as star systems, planets, atoms, cells and quanta. Multiple tones resound simultaneously as harmonious chords of information, driving energy to combine into patterns of dimensionality.

For example, consider the rhythmic pulse of the human heart and Earth's fundamental frequencies—what drives them? What are they responding to? Why is Earth changing its rotation at present? Moreover, why does Earth rotate at all? What is driving the motion of the 20 tectonic plates over the surface of the planet? The answer to all of these is one in the same: the "tones" of creation. The vibration that allows the silica of Earth's crust and carbon of our bodies to align as tetrahedral bonding patterns is the same song. In the only words available during their time, the ancients reminded us of these relationships. Kept alive within the sacred texts of the mystery schools, the relationship between sound and form is a fundamental key in understanding creation.

In the beginning was the Word, and the Word was with God, and the Word was God. He was in the beginning with God; all things were made through him, and without him was not anything made that was made. In him was life, and the life was the light of men.

THE HOLY BIBLE, JOHN 1:1

The "word" was the sound or vibration that initiated creation. Notice that the sound simultaneously was *with* God and also that the sound *was* God. Sanskrit traditions actually offer us the memory of the first sound that pulsed through the undifferentiated "soup" of potential creation; the seed syllable of "OM" (Fig. 24).

Figure 24 *The Sanskrit seed syllable of the original vibration "OM."*

In the great beginning there grew the first cause, that brought into being all that exists. Thou, thyself, art the effect of causation, and in turn are the cause of yet other effects.

THE EMERALD TABLETS OF THOTH,[11] TABLET 12

The study of Cymatic relationships provides a very good analogy for the crystallization of energy from undifferentiated form into the distinctive patterns that we see as matter. The patterns are morphogenetic fields of geometry that are supported within specific wave-ranges of the creation matrix, as defined by their frequency. Just as predictable patterns formed within specific ranges of vibration above, specific geometries are supported within specific portions of creation's matrices.

Platonic Solids: The Codes of the Creator

All patterns of three dimensional creation, including the human form, resolve to energy bonds resulting from one, or some combination of, five simple forms.

These five forms have been the subject of scrutiny and debate for centuries. Religions have developed around the understandings of these forms; mystery schools devoted themselves to preserving the information for the use of future generations. The science of alchemy, often associated with the changing of lead into gold, is rooted in the study of these forms. Alchemists were less concerned with obtaining the actual metals, as with the transformation that the metals experienced in moving from one expression to another. Through this process we experience the mirror of Earth's Shift at present. It is within this transformational shift that the keys to conscious evolution may be found, for these keys provide the map of matter expressing as increasingly complex geometric assemblages of vibration. The basic patterns, literally the geometric codes of creation, are known today as the *Platonic Solids* and physically describe the volumes enclosed by these patterns (Fig. 25).

The term "Platonic" is a reference to the scientist and philosopher Plato and one of his best known works, the *Timaeus*. Within this work, Plato uses the tool of metaphor in the description of a universal cosmology based upon interlocking patterns of geometry. These patterns appear to have been known well in advance of Plato, however, as shown in the archaeological records of forms resulting from these patterns, that may be viewed in the Cairo Museum in Egypt. Within the glass cases are finely crafted models dating to 3,000 years old, models of the forms referred to in the *Timaeus*. Predating these forms are those kept in the Ashmolean Museum at Oxford, England,[12] estimated to have been assembled approximately 1000 years before Plato's time. Though not as finely crafted as the Egyptian forms, these models are quite obviously indications of an awareness of fundamental and geometric nature of the building blocks of creation.

A Platonic Solid may be defined as the surfaces delineating a very special, fully enclosed volume. All lengths defining any portion of the volume are equal, as are the values of all interior angles defining the corners. Conceptually, the solid may be thought of as a single unit cell of form, repeating until it falls back upon itself with adjacent, matching unit cells. Each angle formed by the meeting of the unit cells, and the dimensions of each side of the cell are equal. At present, there are five solids known that meet this criteria. Referred to as the five regular Platonic Solids, they are illustrated below in order of increasing complexity as defined by the number of faces.

Platonic Solid	Number of Faces	Edges	Vertices
1. Tetrahedron	4	6	4
2. Hexahedron (Cube)	6	12	8
3. Octahedron	8	12	6
4. Dodecahedron	12	30	20
5. Icosahedron	20	30	12

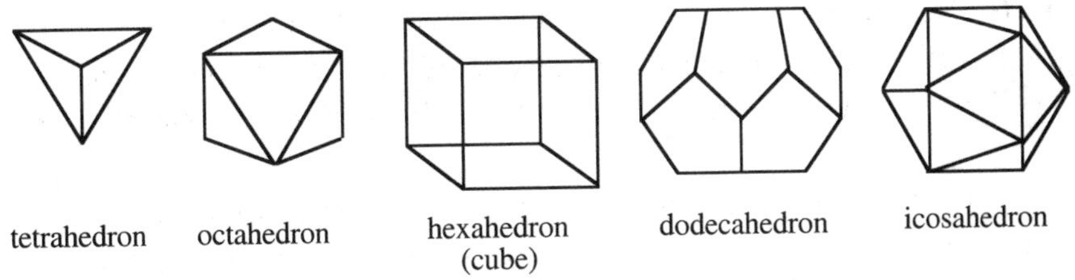

tetrahedron octahedron hexahedron (cube) dodecahedron icosahedron

Figure 25 *Schematic diagram of the five regular Platonic Solids.*

For clarity, the mineral kingdom best exemplifies these natural expressions of energy along predefined guidelines of energy. The naturally occurring structure of the crystal externally is an outward expression of the planes of strength and weakness along which matter has aligned internal to the structure. Atoms combining to form the basic building blocks of the crystal as the first unit cell, define the appearance of the crystal for the remainder of its growth. An example of a common mineral-crystal sequence will help to illustrate this concept.

The mineral compound of table salt is represented chemically as sodium and chlorine (NaCl). The crystalline nature of salt, halite, is not readily apparent while it is dissolved in a solution of water, for example. At specific temperatures and pressures, halite may remain in solution indefinitely, assuming that the ratio of water to salt does not change due to evaporation. Under specific temperatures and pressures, sodium and chlorine will begin to combine, as long as atoms of each are available, and form the crystalline structure that is recognized as salt. This structure is based upon the Platonic Solid of the cube. The bonding of atoms along predefined patterns of geometry is referred to as patterns of packing or *geometric coordination*. Coordination describes the number of units (atoms, molecules, cells, etc.) that comprise the arrangement. For example, "2" coordination describes a linear arrangement, with one of the elements residing at a point adjacent to the next, while the elements in "3" coordination will arrange themselves tangent to one another. Packing patterns are expressed very clearly within the mineral kingdom, as geologists use these patterns to describe the bonding of atoms. The five common coordination patterns based upon the Platonic Solids are as follows:

Description	**Geometry**	**Packing**
1. (2) coordination	Linear	
2. (3) coordination	Equal-sided triangle	
3. (4) coordination	Three-dimensional triangle	
4. (6) coordination	Two four-sided pyramids	
5. (8) coordination	Cube	

Based upon these fundamental forms, geologists have developed a system of crystalline classification for all known minerals. The classification is divided into seven separate crystal systems based upon considerations of symmetry and external expression. The systems are further divided into 32 distinct classes (not pictured), with each crystal resulting from a unique, external geometry, reflecting the ordering and arrangement of atomic patterns. All relate to, or are derived from, one or some combination, of the Platonic Forms.

Crystal System
Triclinic
Monoclinic
Orthorhombic
Tetragonal
Hexagonal (Rhombohedral Division)
Hexagonal (Hexagonal division)
Isometric

The seven systems of crystalline formation. (Adapted from nomenclature proposed by Paul Heinrich Groth, 1895, from the *Manual of Mineralogy* 19th edition. Hurlbut and Klein.[13])

CREATION: INTERLOCKING GRIDS OF SOUND AND FORM

In the discussion of grids, matrices and frequency, the concept of hierarchical grids was developed. These subtle structures of energy support lateral ranges of vibration or frequency. Assembling two or more grids vertically provides a matrix, allowing the energy of one grid to transition into another, above or below. Also from the previous discussion, the relationship between frequency and geometry was identified as a two-way exchange of information. The knowledge of the Platonic Solids and the dynamics of geometry and frequency allow the grid/matrix concept to now go one step further.

In addition to supporting specific ranges of frequency, each grid supports one, or some range, of geometric patterns. The morphology of the grid is based upon the relationship between vibration and form. In other words, specific sound produces predictable and repeatable form. There are grids that support only the form of the tetrahedron; that is, the vibration within that particular grid, *the frequency,* will only produce specific geometric patterns within that grid. Descending to the next range of grids, each of these adheres to the same codes of creation, supporting only specific form. When a portion of creation is viewed as solid, it is actually the logical mind generalizing and averaging information that is occurring upon multiple levels of creation simultaneously (Fig. 26).

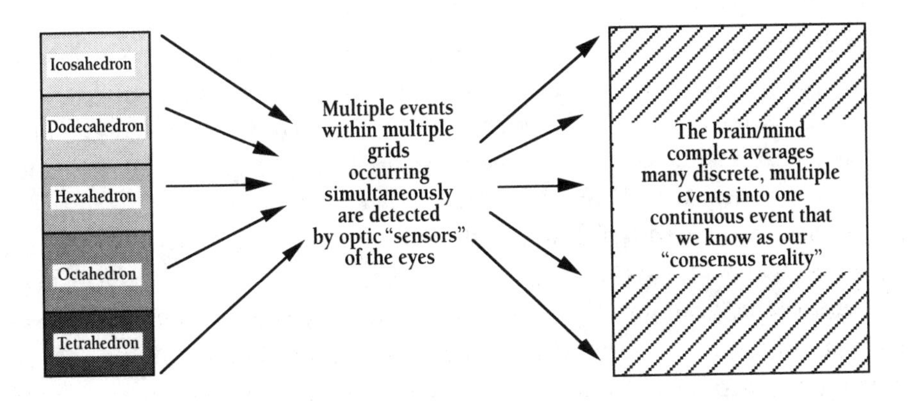

Figure 26 *A conceptual portion of creation. Multiple events occur simultaneously within discrete grid structures. Our mind generalizes the many events into a single continuous reality.*

The nature of the matrices and grids supporting Earth at this particular time may be thought of as a sustained vibration, tones creating stable, reliable "standing wave" patterns over thousands of years. These tones are analogous to the tones played through the stereo speakers in the previous demonstration. Similarly, the grid and matrix pattern of creation may be viewed as analogous to the forms supported by specific ranges of sound.

Physicists now speak of creation as nonsolid and noncontinuous. The world of quantum mechanics has demonstrated that creation is actually expressing in very rapid, short bursts of light. These pulsations are called *quanta* and occur so quickly that, although the eye is capable, the mind-brain complex does not discern between each burst. Instead, the pulses are averaged together into what is perceived as one continuous event, termed creation. In much the same manner, this pulsed light is a glimpse into much more than a single grid of information within the creation matrix. From this perspective, light is actually our mind-brain complex averaging together multiple interlocking grids of geometry, modulated upon radiant energy (light), into a single, continuous event expressed as the continuity of "reality." Through this insight, life may be viewed as a series of very rapid events occurring upon multiple levels of the creation matrix. In addition to being knowable, these levels are navigable.

THE *FLOWER OF LIFE*: THE MATRIX OF CREATION

90 miles from the city of Luxor on the west bank of the Egyptian Nile lies a temple complex that is unlike any temple at any other location upon Earth. Predating the ancient remains of the temple overlying it by several thousand years, the architectural style of this site is unique to that of any site, found anywhere in the world. The closest example of even a similar style is that found upon the Giza complex in Cairo, now being dated geologically at Pre-Ice Age. This site is that of the temple dedicated to resurrection through Osiris; the Osirion.*

Lying below and adjacent to the Seti I complexes in Abydos, presently dated as one of the oldest complexes in Egypt at 6,000 B.P., proximity alone dates the Osirion as earlier, with portions of it actually laying beneath the Seti I site (Fig. 27). The Osirion is constructed of massive granite slabs held together by "tongue and groove" construction, with the walls fitting so precisely that no mortar was required. Originating in the quarries at Aswan, many miles to the south, the mystery remains as to how these monoliths were transported to their present location in Abydos. The indications are that the Osirion Temple is actually much older than currently acknowledged, and provided an environment for the safekeeping and dissemination of a very ancient, extremely sophisticated and comprehensive body of knowledge with direct relevance to the present. In modern terms, we would refer to this site as a mystery school, situated within a hall of records, developed prior to the close of the last subcycle, between 10,500 and 13,000 years ago.The "Codes of Creation," embodied upon the walls of the

* The Osirion is also referred to as the "Temple of Resurrection." Egyptian legend refers to the Osirion as the site where Osiris' wife, Isis, brought the severed portions of her husband's body to be reassembled and imbued once again with life.

Osirion, are keys to understanding our bodies, the continuity of birth, life, death, resurrection and the significance of the "Shift." The codes still work!

Since its discovery, controversy over the message of the Osirion has raged. Torn between the categorization of "science" and "religion," it was neither separately, and both combined. The information preserved upon the walls of this site offer significant clues into the study of the most profound, universally applicable body of information available to humankind at any time in history. The "codes" embodied within the walls of the Osirion are just as valid today as they were 8,000 years ago and, perhaps even more relevant, as the close of the cycle rapidly approaches.

Figure 27 *Interior view of the temple of the Osirion. Columns with the illustration of the* Flower of Life *and the derivatives are to the left.* Author's note: *The Osirion is being inundated by rising ground water and is not open to the public at this time. The rise in the ground water appears to be the result of the "High" dam in Aswan. As of January, 1996, there appear to be plans to stabilize and restore this unique site.*

Inside of the load-bearing columns supporting the massive granite slab ceiling, upon the walls, are a series of patterns appearing as varied groupings of interlocking circles, referenced as "flowers." The arc of each circle bisects the circle adjacent to it at exactly the one half point, passing through the center of that circle. The net effect is that each of the circles are related, having the same diameter, with each edge sharing the space of one half of that of its neighboring circle (Fig. 28). Each complete flower is composed of 19 circles with the entire pattern enclosed within two nested, concentric rings. The rings, tangent to the outermost circles, are referenced in the science of Sacred Geometry as the *zona pelucida*, a Latin term for the inner and outer surfaces of the human ovum. Within the resonant symbol of the flower, the *zona pelucida* is a metaphor mirroring a reality; the codes of all possibility exist holographically within each cell of the human body as dormant potential. Life, experience through the tools of choice and free will, are the activators of these potential realities responding to the inertia of creation focused through the resonator of the body. It appears to be the very experiences of life itself that determine which program patterns are activated. The most sacred aspect of Sacred Geometry, the science of being, is the science of consciously activating the codes to provide life patterns of "will." These are our chosen experiences.

Figure 28 *Illustration of the* Flower of Life *found in the Osirion Temple complex 90 miles West of Luxor, Egypt.*

How the sacred patterns were applied to the stone surface is a mystery unto itself. These beautiful geometric codes have not been etched, painted or carved into the hard rose-granite walls—*they are literally "flash burned" into the stone through a process that is not understood today*. This ancient code is referred to throughout *The Emerald Tablets of Thoth* as the *Flower of Life*.

> *Deep in Earth's heart lies the flower, the source of the Spirit that binds all in its form. For know ye that the Earth is living in body as thou art alive in thine own form. The* Flower of Life *is as thine own place of Spirit, and streams forth through the Earth as thine flows through thy form.*
>
> THE EMERALD TABLETS OF THOTH,[14] TABLET 13

In recent years, the codes contained within the *Flower of Life* have been "resurrected" and decoded as a course of study in the Western world. Through the efforts of a very gifted teacher, Drunvalo Melchizedek, the derivation, message and the meaning of the *Flower of Life* have become reincorporated into the teachings of ancient mysteries through experience, knowledge and memory. The foundation of Drunvalo Melchizedek's work is the awareness and use of "crystalline" fields of energy-information surrounding the human body. These fields are referred to in the ancient texts as fields of Light, Spirit, Body or *Mer-Ka-Ba*.[15]

Various derivations of the *Flower of Life* have become recognized throughout the world as the *Tree of Life* and the *Seed of Life*. Within these patterns are the mathematical sequences, literally the codes of creation, that describe the geometry of light interacting as genetic material within each cell of the human body. In recent months, this author's research suggests that the very manner in which the genetic code arranges itself within the DNA molecule is governed by the *Flower of Life*. The overall length, branching patterns, ratios of branches to the "trunk," even the angles of branches themselves within the essential amino acids are within the limits set forth through the intersection patterns within the Flower (Fig. 29).

Figure 29 *The* Flower of Life, *as a template for three of the essential amino acids.*

The *Flower of Life* details coordinates of the morphogenetic patterns for each of the Platonic Solids: the biological programs orchestrating the direction and growth rate of life; ratios of males to females in unregulated populations; the dendritic branching patterns of trees, plants, root systems, electrical discharges in the atmosphere, proportions of the human body and, especially significant to the mastery of the physical experience, proportions of geometric fields of energy radiating within, and beyond, the human body to create the "nests" of energy supporting our third dimensional experience.

Within this pattern may be found the coded basis for nearly every religious belief, sacred order, metaphysical law and subset of science that has been structured, both ancient and modern. The reference to modern science as a subset is intended as an acknowledgment to the validity of "science," as well as its state of incompleteness. The wisdom of Western technology is based upon partial understandings, distorted by virtue of their incompleteness. The "laws" of physics may be viewed as localized "truths"; they are descriptions of observed patterns of energy within limited parameters of experience. The universal law that applies throughout all of creation, is the law of expressive energy:

As you change the frequency of energy,
you change the patterns through which the energy expresses.

With the law may be found the key to decipher the codes of creation through the vocabulary of the sacred language and observe the patterns of truth for yourself.

The Judeo–Christian traditions, Hinduism, Buddhism, Taoism, the orders of the Red Hand, Golden Dawn, Emerald Cross, Amethystine Cross, Free Masons, the Hebrew Kabbala—all of these doctrines are rooted in the knowledge of fundamental relationship between humankind, the Creator(s) and the world of humankind. Within this path, the study is based upon the vocabulary of a very ancient body of information; that of Sacred Science or more precisely, Sacred Geometry. The "geometric" aspect of Sacred Geometry does not necessarily imply a left-brain, masculine and technologically oriented understanding of the codes of creation—although it certainly may, for those who choose to approach this course of study through those particular "lenses" of expression. The teachings of the ancient mystery schools are offered as a balance of logical left and intuitive right-brain knowledge. Both forms of knowledge are experiential and necessary. Each is exact knowledge of an exact science, whole and complete unto itself.

> *Great healers, men of divine realization, do not cure by chance but by exact knowledge.*
> PARAMAHANSA YOGANANDA

THE *FLOWER OF LIFE*: TEMPLATE OF GOD

Conceptually, the discussion of grids and matrices within this text is accurate. All of creation in general, and our bodies specifically, exist by virtue of matter structurally bound to the framework of grids and matrices supporting patterns of creation. The description of the matrix framework however, was not completely accurate. For simplicity of illustration, the conceptual grids were constructed of cubes and rectangles, allowing the reader an opportunity to focus upon the concepts at hand. The idea of grid cells based upon the square and rectangle is much easier to grasp as much of our modern technology is based upon the geometry of the square. While the concept is accurate, the actual expression of the concept is not.

The matrix of creation, the underlying structure of all form, guiding its propagation and very existence, does exist as a series of interlocking, hierarchical grids. These grids, however, are not cubic or rectangular in nature. Rather, they are based upon spheres. Specifically spheres within spheres; the *Flower of Life*! It is upon this framework and the intersection planes created through interlocking *sphere-cells* that the energy-information-light is directed, translated and propagated throughout creation.

The Flower of Life *provides a finite model,*
of what is essentially an infinite matrix underlying the fabric of creation.

The Emerald Tablets of Thoth describe the standing wave "lattice" of pure life force, the *Flower of Life*, as radiating from a specific point deep within the Earth, an ancient temple site remaining active, though now buried in what was known as the Halls of

Amenti. Through this radiance of pure *prana*, form becomes form and life yields life. It is to this radiant information that Thoth attributes the patterns of creation.

> *Deep in the Halls of Life grew a flower, flaming, expanding, driving backward the night. Placed in the center a ray of great potence, Life giving, Light giving, filling with power all who came near it.*
>
> THE EMERALD TABLETS OF THOTH,[16] TABLET 2

> *Deep in Earth's heart lies the flower, the source of the Spirit that binds all in its form. For know ye that the Earth is living in body as thou art alive in thine own formed form. The* Flower of Life *is as thine own place of Spirit, and streams through the Earth as thine flows through thy form. Giving of life to the Earth and its children, renewing the Spirit from form unto form. This is the Spirit that is form of thy body, shaping and molding into its form.*
>
> THE EMERALD TABLETS OF THOTH,[17] TABLET 13

In the 20th century, a similar premise has been reintroduced by the noted physicist and author Max Planck in his reference to a *force* that appears to hold the particles of the atom together, adhering to an apparent framework of intelligence that he references as "*mind.*" In accepting the Nobel Prize for his study of the atom he made both an interesting and remarkable statement in his speech:

> *As a man who has devoted his whole life to the most clear headed science, to the study of matter, I can tell you as the result of my research about the atoms this much:*
>
> *"There is no matter as such!"*
>
> *All matter originates and exists only by virtue of a force which brings the particles of an atom to vibration and holds this most minute solar system of the atom together.…We must assume behind this force the existence of a conscious and intelligent mind. This mind is the matrix of all matter.*[18]

What is this *mind*—this creative intelligence that exerts a force throughout all aspects of creation? How is it structured within our matrix model of creation? In the beginning the "Word" provided the foundation upon which all subsequent wave patterns would be built; the fundamental standing wave pattern determining the "laws" and "guiding" the crystallization of all forms of vibration. *That fundamental vibration is the* Flower of Life. In two dimensions appearing as circles within circles, the temple walls are actually modeling the three dimensional expression of spheres within spheres; spherical matrices. Each circle within the pattern is whole and complete unto itself while functioning as an integral portion of the overall pattern. Within each sphere, the pattern repeats itself again and within each of those spheres the pattern is reproduced yet again, and again and again. The information within these patterns may be considered as holographic in nature. Within the holographic process, each individual cell has all of the information necessary to initiate and replicate the entire whole once again (Fig. 30).

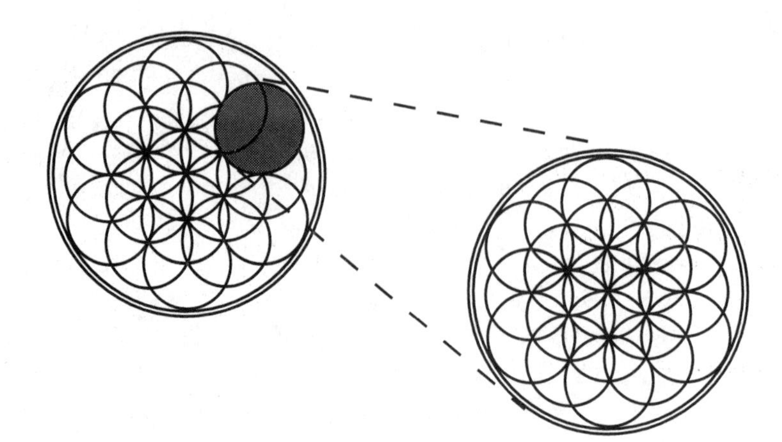

Figure 30 *Illustration of the holographic nature of the energy underlying each aspect of creation. Each "cell" is complete unto itself while containing all of the information required to replicate the whole.*

Each cell is whole and complete unto itself while, at the same time, forming one component of something much more than it may ever be alone. The matrix of the human body is a very familiar example of a holographic process. Within every cell of every organ, all tissue, bone and fluid of each individual, lies a complex molecule that has been the subject of intense scrutiny for decades, the deoxyribonucleic acid molecule or DNA. This twin strand of genetic material is coded with all of the information required to generate a complete replica of the individual. The cell is whole and complete unto itself; a full compliment of information. The same cell is also one part of something much larger; the physical body. As part of the body, that one cell is able to accomplish much more than is possible as a single free-floating cell. The human body is a holographic matrix of cellular information, the full compliment of which is stored within the genetic material of each cell.

It is through the knowledge of an underlying matrix, the standing wave pattern of the *Flower of Life*, that the mechanism of creation becomes apparent. Energy-information-light originates within a source of consciousness and is propagated from this focal point through the "conduit" of a multidimensional matrix; spheres within spheres. The planes of intersection within each of the spheres become the framework along which matter is inclined to crystallize, as it descends into more dense levels of the creation matrix. It is within these intersection points that the true nature of creation may be glimpsed. For it is at the intersection planes of the spheres that the patterns of creation begin to coalesce.

Earlier in this section it was stated that everything that is built, created, uncreated, physically felt or touched within this world is what it is because energy has congealed following some combination of bonding patterns described by the Platonic Solids. How does the energy "know" where to form and how to bond? To understand the answer to this is to understand the morphogenetic fields, or blueprints, of creation. Figure 28 shows the *Flower of Life* the way it is depicted upon the temple walls of the Osirion; a finite representation of 19 circles enclosed within two concentric rings surrounding the entire model.

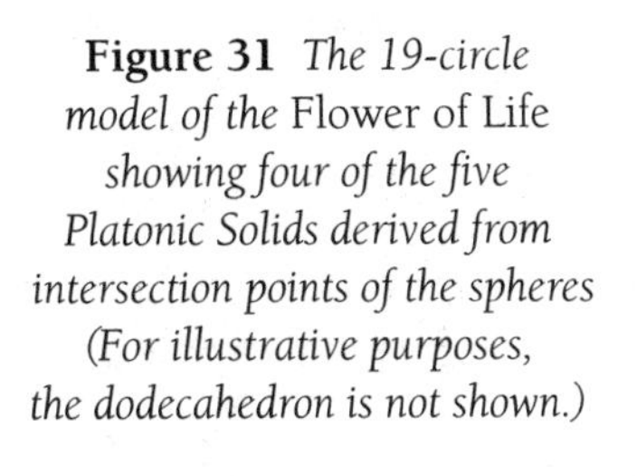

Figure 31 *The 19-circle model of the* Flower of Life *showing four of the five Platonic Solids derived from intersection points of the spheres (For illustrative purposes, the dodecahedron is not shown.)*

The intersection of circles within the *Flower of Life* produces a series of points that may appear random at first. Closer inspection will reveal that the lines connecting the points depict in two dimensions the forms that underlie the bonding patterns of nature (Fig. 31). These patterns are the five Platonic Solids. As information-energy-light cascades through the matrix of creation, it preferentially aligns itself with the unique patterns of expression within those parameters. Light slowing down to the frequencies that we consider to be those of third dimension crystallizes as the only forms that it knows within those parameters. The study of these patterns is the sacred study of the relationship between form (geometry) and vibration (frequency) within the parameters of a given condition. We call that condition dimensionality. In this knowledge the human relationship to creation, Creator and ultimate transition from one expression of being into another, take on new meanings. The mechanism underlying the process is demystified as the roots of religion, beliefs and distortions become apparent.

Equally apparent is the path illuminating the key to our mysteries through the language of creation. This path may be found by looking within. Each cell within our body mirrors a holographic process described as the *Flower of Life*. This is the reason that this pattern is so familiar. Even if it is not known consciously we live the pattern in every moment of life. The *Flower of Life* is the basis for the teachings of the ancient mystery schools; another language, another perspective, an additional tool with which we may know ourselves.

> *In the beginning was the Word, and the Word was with God, and the Word was God.*
> THE HOLY BIBLE, JOHN 1:1

The Biblical texts remind us that the "Word" was not only *with* God, but the word *was* God. The word, or tone, was spoken into the soup of potential. This potential is the void referred to by the Egyptians as "Nun." From the study of Cymatics it may be seen that vibration of the word will produce a pattern within the given medium. Into the void of Nun was sounded the Word providing the fundamental patterns of vibration upon which all other vibration would be built and expressed. "In the beginning was the Word"—the *Flower of Life*—the standing wave pattern of life-force-potential that produced the base vibration. This vibration became the framework upon which all of creation would be formed. In the beginning, and still existing, was-is the fundamental intelligence underlying the entire fabric of creation: The *God* Force.

Rooted deeply within the creation stories of many indigenous traditions is the premise that the people of each particular clan are fashioned in the image of their Creator. They make a distinction between this Creator and a universal God. The ancient records of these people acknowledge the existence of many creators responsible for the creation of many clans. Each clan however, acknowledges an infinite intelligence that underlies all of the universe, including the creators of the creators. What does this intelligence, the force field of God, look like? How may we know of this force within the parameters of our third dimensional mental and emotional constructs? Perhaps our study of the relationship between vibration and form may provide a language from which to express this relationship.

The highest expression of geometric form attainable within the parameters of our Earth experience is the sphere. The form of the sphere represents both the most complex, as well as the most generalized geometry possible. The sacred sciences remind us that rather than a single sphere, we see this pattern holographically as interrelated spheres within spheres. Ancient manuscripts reveal this relationship as the *Flower of Life*. It is through the tuned and resonant symbol of the *Flower of Life* that we are offered a visual model from which we may logically know of the patterns representing the intelligence of creation. From this perspective of vibration, form and experience, the *Flower of Life* represents the pattern of God.

THE *SEED OF LIFE*: A MODEL OF LIVING KNOWLEDGE

The *Flower of Life* represents much more than the codes of our creative matrix. As Drunvalo states, the *Flower of Life* is

> "...the sum of all that may be logically known, rather than emotionally felt, within our world."

Reflected in the derivation of the *Flower of Life*, from the *Seed of Life*, is the metaphor for the force of knowledge as living information. This force, merged into a preexisting field of knowledge, yields something greater through the union of the two than either could have offered alone. Within the mystery schools, the *Flower of Life* unfolds from the *Seed of Life* and the Seed from the Cosmic Void—the original nothingness of the undifferentiated potential. The following series of illustrations depict this unfolding through metaphor.

Ancient Egyptian mystery schools used the glyph of the eye of Horus to represent individual, group and mass consciousness. Specifically, the right eye of Horus is used to indicate this undifferentiated state of awareness. In recent years, the eye of Horus has been recognized as being an entire informational system encoded within itself. The keys to this system appear to be based upon the relative proportions of the glyph's components, and are explored within a unique work dedicated to this system, *The Brilliant Eye* by Alexander Joseph.[19]

Figure 32
The Eye of Horus as the Egyptian glyph of consciousness.

The sequence of the first seven iterations of conscious expansion is adapted, and expanded upon with permission, from Drunvalo and the *Flower of Life Workshops*[20] in an effort to illustrate the metaphor of creation that begins within the womb at conception.

In the beginning there was the void; nothing and everything existing simultaneously as undifferentiated potential. This was consciousness unknowing of itself as an individual, as a group; consciousness had no frame of reference—it just "was."

At some point, after some period of time, consciousness began to develop an awareness of itself as "something" and in doing so also began to develop the awareness that it was existing "somewhere." Still without a frame of reference, the directional terms of up, down, front, back, left, right were meaningless. At this point, however, consciousness *knew* that it *was* (Fig. 32).

After experiencing in this state for some period of time, it could have been milliseconds or millennia, consciousness began to develop a frame of reference and chose to define a world of knowing for itself. This was accomplished through choosing an arbitrary distance of *knowing* and, using this distance as a reference, projected its awareness one *knowing distance* above, one *knowing distance* below, one *knowing distance* to the front and back and to the left and the right, surrounding itself within a motionless field of experience. This field was static in all directions. Consciousness then began to *rotate* about each of these three axes simultaneously, and in doing so provided the mechanism for its zone of experience to shift from a static to a dynamic field of knowing. This rotation became the primal sphere of experience (Fig. 33).

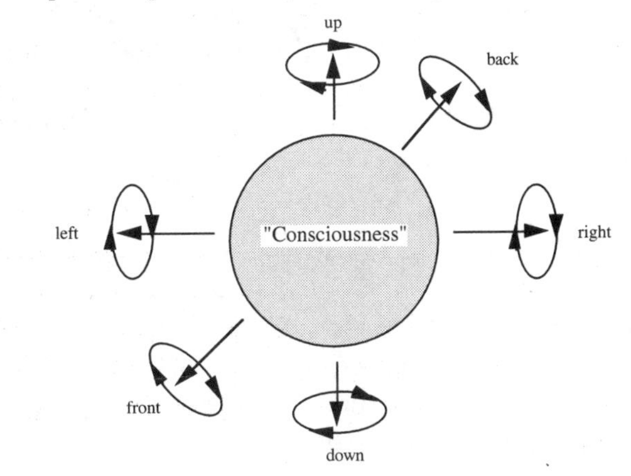

Figure 33 *Consciousness became aware of itself as "something" and began rotating about the three axes that defined its sphere of knowing. In doing so, the sphere became the frame of reference within which to define itself.*

With the creation of the sphere, consciousness had defined for itself a world of safety and knowledge, remaining unaware of any events occurring beyond the boundary of the sphere. The space within the sphere was the only world that consciousness could know at that time, it's own sphere of awareness.

At some point consciousness chose to expand the limits of its experience beyond the bounds of what it knew, to move into the unknown and integrate the unknown into the whole of its experience. This was accomplished as consciousness moved to the edge of its known world (in this example the top of the sphere), and from this vantage point projected its awareness one *knowing distance* once again into the void of the unknown. At this point, the metaphor becomes interesting with relevance. Consciousness did precisely what we do today as we move into the unknown spheres of our life experience. We begin by taking stock of our known world—our tools and experiences—and with these tools in hand move to the very edge of our world as we have defined it. We push the boundaries of what we have been conditioned to accept as our knowing world through the wisdom of our experience. From the very edge of our known world we step off into the "void of the unknown" with the only tools that we could possibly take with us, that which we know, and begin to define a new "sphere" of experience within that which was previously unknown.

Whether it is a new career, relationship, physical or spiritual journey the path is the same. We project our sphere of awareness into the unknown, that which is not yet of our experience, and explore the unknown through the lens of what we know to be true.

As consciousness performed this subtle yet powerful act of exploration, something wonderful and awesome occurred that would forever change the manner in which consciousness related to experience.

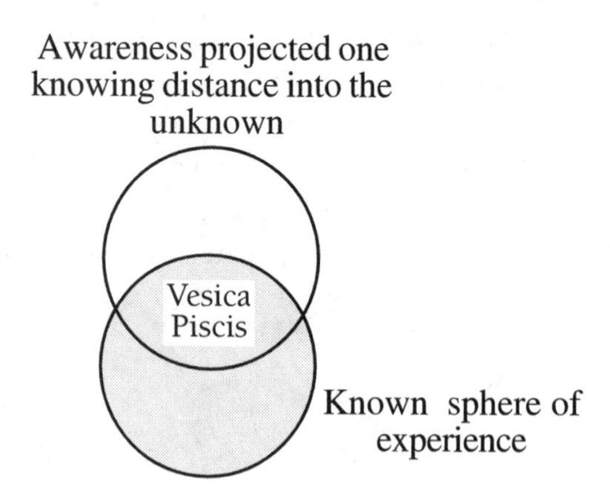

Figure 34 *The first projection of the "known" into the "unknown," creating the* Vesica Piscis.

Figure 34 illustrates the act of projecting the known into the unknown from a vantage point of the edge of the sphere. As the two perfect spheres are formed in close proximity (each contains half the diameter of the other), a zone of commonality is created in the overlap. In the science of Sacred Geometry, this zone is referred to as the *Vesica Piscis*. The form of the vesica was used as both the Egyptian glyph for the "mouth" as well as for the Creator. Additionally, this glyph is also very similar to the Mayan glyph for zero, associated with our galaxy of the Milky Way. Within this zone of the *Vesica Pisces*, something truly magic began to occur. As consciousness experienced within the void, the entire world that it knew was within the context of that sphere, and only that sphere. All was in perfect order, perfect balance and perfect harmony, and perfectly uneventful! Energy was homogeneous throughout. As the second sphere was generated, however, the perfect order of one sphere of experience is not exactly equal to the perfect order of the second. The zone of overlap is the zone of tension or stress caused as the two spheres seek resolution and attempt to balance one another. It was this tension that provided the inertia to drive consciousness toward balance—*the tension of disorder seeking order.* Some "thing" began to happen. There was movement; motion. Energy within the two spheres of creation was being driven to achieve balance.

Energy was seeking to resolve difference in an effort to regain the harmony and return to the One.

At some point consciousness chose to expand its known world and went through a similar process; moving to the edge of its known world and projecting into the void of the unknown from its known experience. This time, however, the projection was from the intersection point of the two existing spheres of experience. You begin to see a pattern developing (Fig. 35). The projections continue again and again and again until consciousness reaches the point where it began: the return.

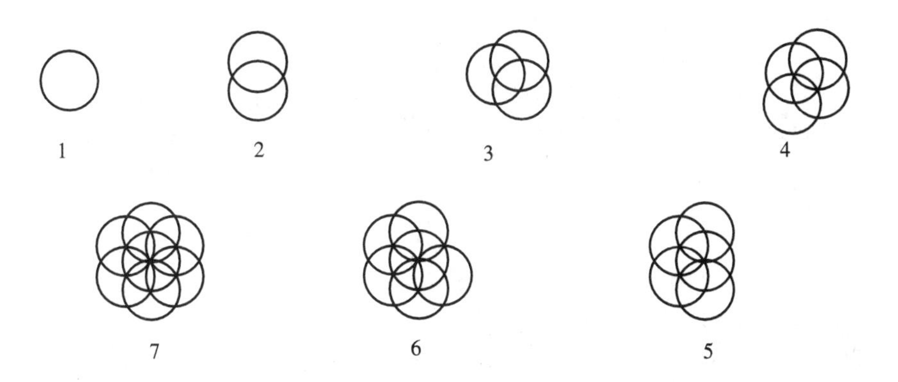

Figure 35 *Consciousness expanding simultaneously, from the intersection points of the spheres, into the void. The first seven geometric expansions yield the* Seed of Life *(pattern number "7").*

At this point consciousness had completed its journey into the unknown at that level of experience. To continue it had to expand from the intersection points at the outer perimeter of the newly defined world. In completing the journey within one particular level of experience, consciousness had just described a second very powerful resonant symbol, that of the *Seed of Life* (Fig. 36).

Figure 36 *The first seven expansive iterations of "spherical consciousness" produce the* Seed of Life.

Though highly esoteric, the *Seed of Life* may serve as the metaphor for a many and varied types of experience. Consider the seven interlocking spheres for example. In the Biblical traditions they may represent the seven days of creation, with each sphere representing an entire grid-matrix relationship within the early stages of creation. A physicist examining the "Big Bang" theory of energy exploding into a preexisting system to create a new universe, may view the *Seed of Life* as the mechanism allowing new energy entry into the existing system. The seven notes of the musical scale; do, re, mi, fa, so, la, ti and with do the return to the first or the beginning. In the book of *The Emerald Tablets,* Thoth refers to the *Seed of Life* mystically as a "key" to the secrets of knowledge within the Great Pyramid and the entrance into the Halls of Amenti.

> *Follow the KEY I leave behind me, seek and the doorway to Life shall be thine. Seek thou in my pyramid, deep in the passage that ends in a wall, use thou the KEY of the SEVEN...* THE EMERALD TABLETS OF THOTH,[21] TABLET 4

As the seed expands into the flower, it does so by simultaneously moving to the intersection points of its present spheres of knowing and projecting one *known length of wisdom* at a time into the unknown until it meets itself as in the musical scale, then continues again expanding to a new octave of experience. The human cell, progressing through its subcycles of division, follows seven recognized stages before it becomes two. The *Seed of Life* may be thought of as the model for the expansion of new energy-information-light into the media of a preexisting intelligent form of energy-information-light. We live this model physiologically, from the instant of conception, as well as emotionally as our sphere of awareness expands into the unknown.

The truly sacred aspect of Sacred Geometry is the relationship between form, vibration and ourselves. From the instant of conception, we develop following specific, repeatable and predefined patterns of form. These patterns repeat themselves through the symmetry of life expressing geometry over and over throughout the developmental stages of the fetus.

Our bodies are encoded through three primary mathematical sequences. These programs may be considered as literal blueprints of light, bringing to us the codes of creation. One of the most powerful of the codes is that of the *binary sequence*, a portion of which is illustrated in Table 6.

The Binary Sequence

1, 2, 4, 8, 16, 32, 64, 128, 256, 512, 1024, 2048...

Table 6 *The first 12 elements of the binary sequence.*

Each successive term in the sequence is found by doubling the previous term. The binary sequence may also be one of the most familiar, as the memory of modern computers is based upon this series of numbers. There is another expression of the binary code that may be even more familiar. We recognize this code on a genetic level. Examining the genetic code for carbon-based life, we find that there are 64 possible combinations of the Carbon, Oxygen, Hydrogen and Nitrogen that allow us the expression of human. Represented as four chemical bases located at three positions, simple math illustrates the binary possibility; $4^3 = 64$, the *seventh* member in the binary sequence and completing number of the musical scale. The real power of this sequence becomes apparent in consideration of the large values that may be reached in a relatively few number of operations.

We make use of this code to reproduce our body's cells. Research indicates that the average human body contains approximately one quadrillion cells that seek replication many times each day. A linear replacement, one at a time, would be out of the question. The binary sequence, however, provides the mechanism for the body to accomplish this task multiple times per day. The term for the division of one cell into two complete cells is that of mitotic division. One cell becomes two. Each of those cells divides into two and so on (Fig. 37).

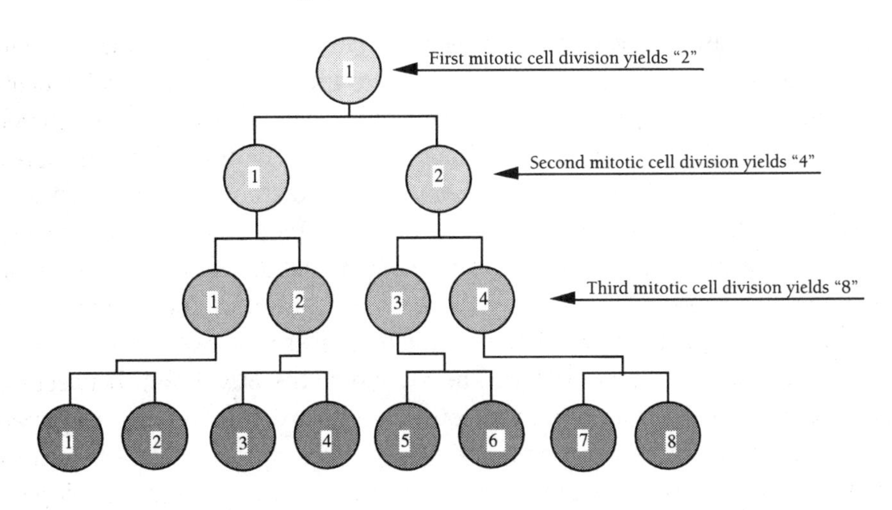

Figure 37 *The binary sequence illustrating mitotic cell division.*

Through only 46 mitotic cell divisions, the goal of 10^{14} power, one quadrillion cells, is reached. It is through the process of the binary sequence that the geometry of the cells becomes apparent. Through the tools of fiber optics and electromicroscopy, the binary sequence of mitotic cell division becomes visible. The single cell becomes two. Each of those cells in turn become two, and so on. Close examination of the cells will illustrate, however, that the cell divisions do not yield random assemblages of cells. Rather, they follow a cyclic pattern of forms, moving from one to the next before cycling through again and again. The drawing of an imaginary line connecting the centers of each of the "spheres" of the cell will determine the pattern of geometry.

The pattern is as follows: The first cell becomes two. As each of these becomes two, the four always arrange themselves in a specific pattern following the geometric form of the simplest Platonic Solid; the tetrahedron (Fig. 38 left). With the next division there are eight cells that have arranged themselves as two *interpenetrating tetrahedrons*, rather than side by side, describing the form referred to in Sacred Geometry as the interlocking or *Star Tetrahedron.*

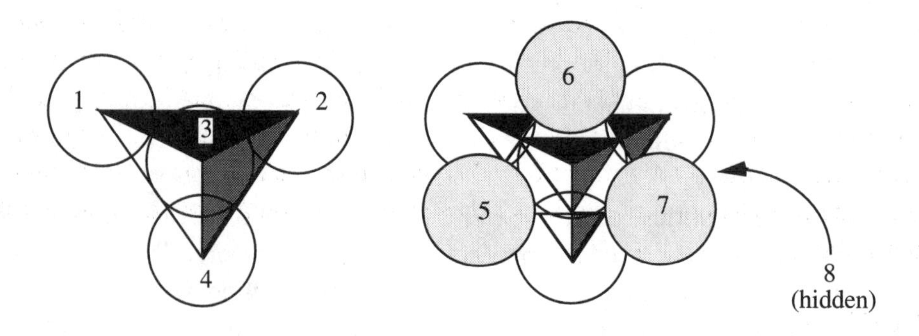

Figure 38 Left: *A schematic illustration of the first two mitotic cell divisions.* Right: *The first three mitotic cell divisions produce the eight cells of the double tetrahedron.*

It is within this pattern of the Star Tetrahedron that we will approach an understanding of the human relationship within itself specifically, and within creation in general. In two dimensions, the Star Tetrahedron appears as a familiar form known as the Star of David, also as Mogen David in the Hebrew faith. A very powerful resonant image, this symbol holds an almost universal familiarity. The six-pointed star represents a form of primal completeness, expressed as a radiant pattern of energy that actually extends beyond the physical body. Whether shown to a Quechua Indian in the Peruvian Andes, a Native American in the desert Southwest, or an Egyptian in a Nile village, all respond to this symbol with a knowing that it is not only meaningful, it is meaningful to them. How could they know? Why is the Star Tetrahedron rooted so deeply into the memory of human consciousness and almost universally recognized as a powerful and meaningful representation of something within?

The patterns of the Star Tetrahedron, as well as that of the *Flower of Life*, are reflections of codes that are deeply etched into each of us at a level much greater than that of conscious knowing. These patterns are mirrors of the language that allowed each of us to emerge into this world, the language of light, form and sound. Representing something far greater than simply stored information, these patterns are literally the geometry of the coded gene. Every human upon Earth not only recognizes these patterns on some level, each *knows* them very well. They *have* to in order to achieve the expression of physical body in the Earth experience.

Each cell continues to divide, following the successive blueprint of the Platonic Solids—Tetrahedron, Octahedron, Hexahedron, Dodecahedron, Icosahedron—cycling through each again and again and again. To provide clarity to the reader, the forms generated through the cellular bonding produce a unique field of radiant information providing the interface between the physical body, and the matrix supporting the body. Beyond the energy of the auric or pranic fields, these are subtle "crystalline" fields of information-energy that resonantly link each cell of the physical body with its individual nonphysical source.

For the purposes of this text, we will stop with the first 2^3 power $(2 \times 2 \times 2)$ mitotic or eight cell divisions. There is, however, an entire and very exact science detailing the presence, access and utilization of all of these fields individually and in combination. A very ancient science, the awareness of these radiant patterns of information is a very empowering body of knowledge that was relegated to the churches and mystery schools approximately three to four thousand years ago. Passed down in an unwritten format to the present day, the ancient knowledge is resurfacing now as we near the completion of the cycle, for those who seek and are willing to accept the responsibility that comes with knowing.

MER-KA-BA
THE VEHICLE OF RESURRECTION/ASCENSION

The Egyptian texts offer a parable from the master, Thoth, that he passed to his students during their course of study:

One becomes two;
Two becomes four;
Four becomes eight and eight is equal to one.

This is Thoth's coded reference to the first eight cell divisions producing the radiant field of information described as the Star Tetrahedron. Why the Star? Why is this form so significant within our process?

To answer the question of "why" in relation to the Star Tetrahedron, we must once again refer to the ancient texts and their reference to a field of energy-information-light radiating outward through the human body from a specific point. The name of this field, interestingly, is found in both the Egyptian and Hebrew vocabularies; the term is that of *Mer-Ka-Ba.* In Egyptian, it translates literally to

Mer = Light
Ka = Spirit
Ba = Body

or the Light-Spirit-Body complex of energy that surrounds each cell of the human body individually as well as the entire body as a composite field. Radiating electrical impulses, this particular form originates from the first eight cells immediately following conception, the cells that form the Star Tetrahedron. The *Mer-Ka-Ba* seed is formed, and remains within a point at the root or first *chakra,* located at the base of the spine. At this point, the first eight cells are fixed, anchored firmly at the geographic center of the human body. The pattern of these cells provides the blueprint for a radiant field of energy that extends through and beyond the physical boundaries of the body. The field appears, and is in proportion to, that illustrated in Figure 38 right, the *Mer-Ka-Ba* field.

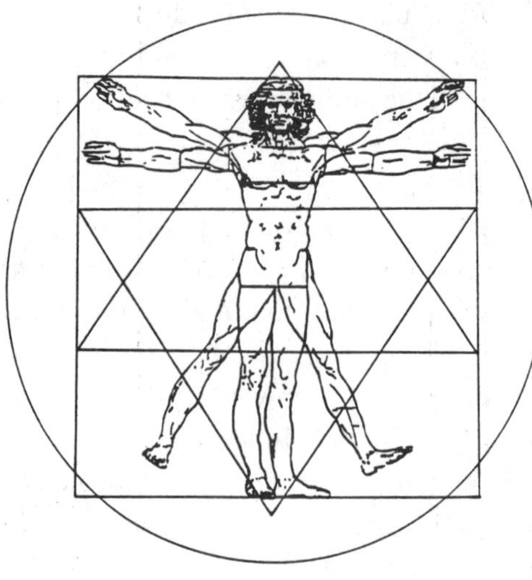

Figure 39
The field of Mer-Ka-Ba *relative to the proportions of the human body.* (After Drunvalo, 1994.)

References to the *Mer-Ka-Ba* have been deleted from much of our open literature. Preserved through the esoteric teachings of the mystery schools, shamanistic initiations and religious doctrines, an awareness of the field of *Mer-Ka-Ba* provides a tool toward the access of creation, the "God force" or the energy of our Creator, directly and consistently. The *Mer-Ka-Ba* may be thought of as a vehicle, in some references, the Light Body or Time-Space Vehicle, capable of transcending the perceived limitations of space, time, space-time, and dimensionality. The key to this understanding is that the vehicle is not something separate from the body, it is an aspect of the body and with specific direction, entrains each cell of the body to a specific and key resonance. In its static phase the field appears as the diagram in Figure 39.

The Star of David is a reminder of this aspect of ourselves, of each individual, a tool to reaffirm and reestablish the connection with greater and more subtle aspects of the God force. Within each individual, through the life experience, the fields tend toward a dynamic phase, that of rotation at specific speeds and ratios. Beyond the scope and capability of the written word, the dynamics of the *Mer-Ka-Ba* are detailed as experience through workshops of the same name.

The dynamic phase (Fig. 40) of the *Mer-Ka-Ba* appears quite different from that of the static (Fig. 39) as the fields begin to assume a flattened, "saucer-like" expression. At key speeds and ratios, the top and bottom of the tetrahedrons begin to move toward one another as the edges of the form expand outward. At these rates of rotation, whole-number multiples of light speed, the extremes of polarity begin to neutralize one another as they move toward a new expression of themselves. This geometric state-of-being may best be described as that of "unity."

The dynamic form of *Mer-Ka-Ba* is also a very powerful resonant symbol, deeply ingrained within the conscious matrix of humankind as the spirals of rotation. The form of *Mer-Ka-Ba* repeats itself as a constant, varying in scale, remaining consistent in ratio of size. From the edge view of the atom, to that of the cell, to that of the solar system, to that of the spiral galaxies, to that of alleged E.T. craft the pattern is the same; the universal field of the human *Mer-Ka-Ba* Light Body.

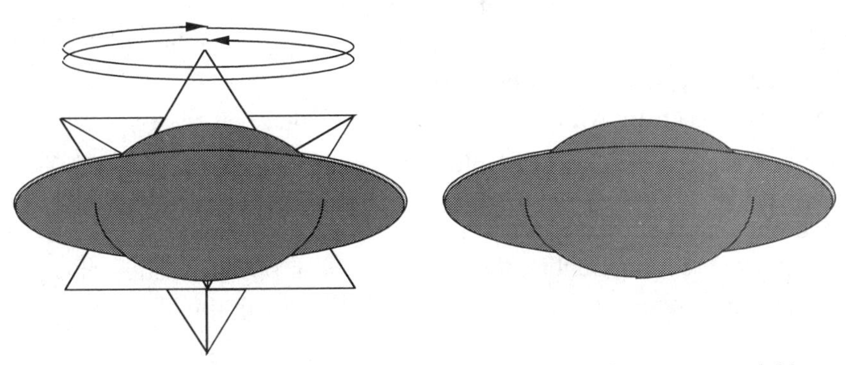

Figure 40 *The dynamic aspect of the Star Tetrahedron, the* Mer-Ka-Ba *field.*

It is the field of *Mer-Ka-Ba* that is described, coded, in the most famous of Leonardo da Vinci's drawings. One page from the notebooks of his mystery school clearly illustrate the proportions of the human figure and their relationship to the *Mer-Ka-Ba* (Fig. 41). The information encoded within this single drawing is so vast that to fully do it justice is beyond the scope of this text. With respect to the *Mer-Ka-Ba* field, however, the focus of the drawing is the frontal form of a man with his arms and legs spread to positions at key angles and enclosed within a circle superimposed upon a square. The angle of the arms is that of the ratio of 5 divided by 2. This is the identical angle of the diagonal from corner to corner within the King's Chamber of the Great Pyramid, an ancient initiation site of *Mer-Ka-Ba* technology.

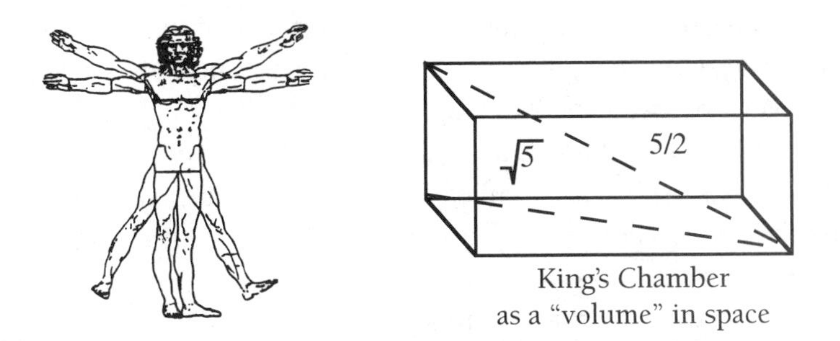

Figure 41 *Relationship between the human proportions depicted in Leonardo's drawing and the geometric relationship of the King's Chamber. The angle of the arms, relative to the horizontal plane of the figure to the left, is the identical angle, in "code," of the diagonals within the chamber to the right.*

The circle surrounding the body in Figure 39 is a perfect circle in two dimensions representing a sphere in three dimensions and is formed as a series of arcs touching the tips of a radiant pattern of information centered at the base of spine, the Star Tetrahedron. The square is a two dimensional representation of the cube in three dimensions and has a very special relationship to that particular circle. The ancient texts refer to the sphere as consciousness perfected, the clearest form of geometry that a human may hope to attain within the Earth experience. Each individual progresses through the grids of experience, successively complex forms of geometry represented by the Platonic Solids, approximating the sphere. As the most complex of the Platonic Forms, the Icosahedron is actually the geometry of perfection for the Earth experience. There are woodcuts and stained glass in European churches depicting the Christ child within a "bubble" of geometric form, that of the Icosahedron.

The goal of each individual is to balance the energy of the physical expression with that of spirit: the cube with the sphere. Mathematically, this balance may be found when the perimeter of the square is equal to the circumference of the circle. In the life experience, the balance of mind, emotion, brain and heart tempered through experience is the key to the long sought after balance between physical and nonphysical.

From those first few seconds of creation through the entire life span and completion of the life, these dynamic fields of light function in direct response to patterns of frequency primarily generated through resonance to Earth's fundamental frequencies. *Primarily* is emphasized here, because each individual has always had the opportunity within themselves to access these fields directly, in the absence of the "auto pilot" mode assumed at birth. This access is generally accomplished UNconsciously through the experience of life itself. The situations and circumstances that an individual masterfully draws into their awareness provides the opportunity to experience and feel differently. The *Mer-Ka-Ba* responds directly to feeling and emotion, as well as thought.

It is through conscious access to the *Mer-Ka-Ba* field that we may *remember* to regulate our physical body's response to external factors—ultraviolet light, viruses and bacteria, environmental toxins—and adapt to rapid shifts within Earth's parameters. All of these are accomplished by vibrating each cell of the mind-body-spirit complex much differently, effectively moving the body out of resonance with one kind of information and into resonance with a higher body of information. Consciously vibrating differently—*regulating the body vibration through thought and feeling*—is where you may discover your greatest power. Life is your key! As with any knowledge, there is a responsibility that comes with the knowing.

PLEASE NOTE: *Conscious access to the* Mer-Ka-Ba *fields is offered and intended from a place of deep love and respect for the gift of the body and life. Any other intent will experience itself, and the consequences of itself, within the fields of the* Mer-Ka-Ba *as it radiates through the body into creation.*

To be very clear with the reader at this point, there are additional morphogenetic fields of information radiating from subsequent patterns of cells, generated immediately following conception. These fields are cyclic and nested following the forms of the five Platonic Solids. Each contains identical ratios of form varying in scale as they project through the hierarchy of creation. These fields may be considered *Mer-Ka-Ba* fields, each a course of study unto itself. There is an entire science, a very exact science based upon the access and use of these patterns individually and as composite fields. When one individual field is accessed, all fields of that form are also accessed through resonance. This is the science of *Mer-Ka-Ba*.

SUMMARY

- The term "Light" is used throughout the ancient texts in reference to broad-spectrum radiant, electromagnetic information; inclusive while not limited to, the visible portion that we are able to detect with the eyes. We are bathed each second of our existence with Light and the patterns of information that are carried by the Light into our bodies at the cellular and genetic level.

- For several thousand years, Earth has been functioning at a base resonant frequency of approximately eight cycles per second (7.8 Hz). In the study of human brain waves, this correlates to a deeply relaxed state of awareness; Earth has essentially been "asleep." As predicted by the Fibonacci series, Earth appears to be moving toward a fundamental vibration of approximately 13 cycles per second, correlating to a waking state of the brain's functioning.

- The underlying fabric of creation may be thought of as two dimensional networks of evenly spaced patterns of wave form (a "standing" wave form). It is along these conduits, defined as grids, that energy moves preferentially. As these networks overlie one another in a "stacked" fashion, the three dimensional aspect of the grids begin to express as matrices.

- Dimensionality may be thought of as varied expressions of matter responding to shifts in vibration. There is a consistent and repeatable pattern that will entrain (guide) the expression of energy in response to specific patterns of sound (frequency).

  Sound yields form...Form yields sound.

  [Geometry <==> Frequency]

- The codes of 3-space creation are based upon one, or some combination of, five fundamental patterns known as the Platonic Solids: Tetrahedron, Octahedron, Hexahedron (cube), Dodecahedron and Icosahedron.

- The *Flower of Life* is an ancient code that embodies the memory of the human form, the continuity of birth, death, life, resurrection and the significance of The Shift of the Ages.

- The *Seed of Life* is the geometric metaphor for the manner in which patterns of knowledge are introduced into a world of preexisting patterns of information. The *Seed of Life* is the precursor to the *Flower of Life*.

- *Mer-Ka-Ba* is both an Egyptian and Hebrew word for the morphogenetic field of energy-information-light that radiates from the human body as the double or Star Tetrahedron. It is this pattern that forms the Star of David. The *Mer-Ka-Ba* is known as the Light Body, Light Body Vehicle, Time-Space Vehicle and the Vehicle of Ascension. It is through the specific ratios and rates of rotation that the fields of *Mer-Ka-Ba* "wrinkle" the grids, allowing the experience of varied dimensionality to consciousness in resonance with the field.

CHAPTER THREE

Interdimensional Circuitry

Our Web of Consciousness

Man is in the process of changing, to forms that are not of this world; grows he in time to the formless, a plane on the cycle above. Know ye, ye must become formless before ye are one with the light.

ADAPTED FROM *THE EMERALD TABLETS OF THOTH*

Our bodies may be considered as a focal point offering a single point of awareness in a multilayered continuum of creation. Our form is *tantrically* linked to multiple aspects of itself, each expressing as a range of frequency, expressing through a nested framework of energy; spheres within spheres. In this respect, our bodies may be viewed as further expression of creation rather than separate and distinct from creation. Through our bodies, we experience a "thicker" portion of creation's matrix. Through our bodies we touch, feel and emote. Our ability to "shift" awareness from one level to another may be considered as our *navigation* of the matrix. Transition through the boundaries between physical and nonphysical is accomplished as the singular act of *vibrating differently.* Through the belief systems born of life experience, we remember to think the thoughts and feel the feelings allowing us to shift. *Our bodies are the focal point*, the resonator, through which wave forms, created high within our creation matrix, are able to coalesce as a form of expression.

The form of our bodies is the tool allowing the outward, giving expression of higher knowledge, as well as the internalized experience of receiving within a polarity world. It is through our bodies that the perceived limitations of our grid-matrix network are overcome. Stated another way, our immortality may be accomplished through the wisdom of Resurrection. Repeatedly within the ancient texts we are reminded that our bodies are a "gift" to be used wisely. With reference to the body as the vessel or temple and the "God force," the intelligence underlying all of creation, Jesus of Nazareth admonished:

> *Do you not know that you are God's temple, and that God's spirit dwells within you? For God's temple is holy, and that temple you are.*
>
> THE HOLY BIBLE, I CORINTHIANS 3:16

Organs of Light: Touching the Invisible

Our bodies may be thought of as a series of distinct, yet related systems of energy. From the subtle geometric patterns of the fields of *Mer-Ka-Ba*, the energy of light is stepped down as coarser gradients of frequency, into the physical body through a system of interfaces referred to in the ancient texts as *chakras* (Fig. 42). The chakra may be defined as the interface point between the physical aspects of the human body and the subtle, nonphysical framework from which the physical is derived.

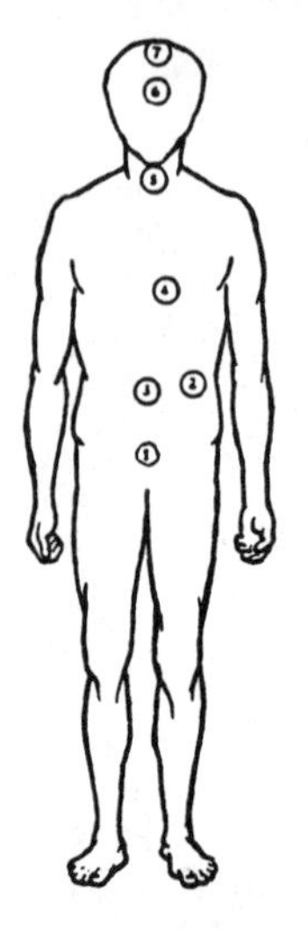

Figure 42 *The present-day view of the chakra system showing frequencies of the seven vortices as follows: 1 = red, 2 = orange, 3 = yellow, 4 = green, 5 = blue, 6 = indigo, 7 = violet.*

These zones are the structural blueprints, serving as conduits for the complement of energy-information that is stepped down through the matrix into the physical expression of that energy. In ancient texts, the chakras are depicted as wheels in motion, specific in color per given location on the body, and lying along a linear pathway somewhere between the Crown chakra, located over the top of the cranium, and the Root chakra, located at the base of the spine.

Within recent years Western researchers have been able to validate the references to these ancient texts through the use of sophisticated, noninvasive scanners that simply "listen" to the body's radiant frequencies and are able to map these systems with precision.* From these scans it is apparent that while the chakra itself appears as a flattened funnel shape, the actual location where the narrow portion of the vortex contacts the human body is a minute pinpoint; less than 1/8th of an inch in diameter! Each chakra rotates at a specific speed that correlates with the frequency of the information to which it is tuned. The vitality of each center, measured as a function of rotation and distance outwardly projected from the body, is a determining factor in the deterioration of the human body. A distinction is made here between the process of *aging* and that of *deterioration.* The aging process is a chronological sequence of events that does not necessarily equate to deterioration. Aging is a forward motion of many discrete experiences, linked together into that which the mind perceives as time. For

*The "Holographic Spectrum Analyzer"™ was developed in the mid-1980s through the efforts of scientist and researcher Robert Dratch. The analyzer has become a primary research tool in the study of interspecies communication, specifically within the Cetacean families.

example, at this very instant you are 20 seconds older than you were 20 seconds ago. During this period of time, have you deteriorated? If so, why? The passage of linear time does not necessarily equate to deterioration over that time. Human aging and human deterioration each are a function of energetic factors of Earth, and our ability to achieve resonance with these factors.

A vast study and science unto itself, for the purposes of this discussion suffice it to say that to the degree our chakra system is maintained at an optimum rotational speed and extended outward from the body, to that degree will the cells dependent upon that vital center for "life force" receive the proper complement of information through the interfacing membrane of the chakra.

While the number of chakras referred to in the ancient texts appears to be accurate, the exact location of the organ is not. Each chakra point is actually located directly over one gland, within a system of glands, that function as transformers in our bodies. The purpose of these transformers is to step down high-frequency information from subtle portions of the grid, to coarser frequencies at manageable points lower in frequency. Used hierarchically within the human energy system, there are seven endocrine glands as shown in Fig.43. The wavelength column in the illustration details the actual vibration of the color associated within the gland-chakra relationship. Measurements are offered in units of Angstrom.

Belief systems over the centuries have varied in terms of the number and location of human energy vortices. Some ancient Chinese systems, for example, indicate four primary vortex points, providing the basis for oriental techniques of acupuncture and acupressure. Modern systems indicate seven to eight chakras and have carried the use of these pathways, also known as meridians, to new levels of sophistication in pioneering modern applications to an ancient, yet sophisticated healing modality; the science of Body Circuits. Each of these sciences address the interconnecting pathways that move energy throughout the body, from chakra point to chakra point.

Though each tradition may recognize a different number of points, there is consistency in each of these systems as to how these points are represented. Regardless of the number of chakras indicated, the designation for a particular chakra at a particular location has the same color reference. Modern systems still recognize the four primary chakras, in addition to at least 21 secondary chakras, and use the same colors to represent these areas. In each of these systems, the first chakra is always red; the third is always yellow; the fourth is always blue and the sixth is always indigo. This consistency reflects the accuracy of the ancient, as well as the modern use of the system.

In viewing the modern eight chakra system (Fig. 42), the colors appear from the root, or first chakra, to the crown or seventh in a specific order. The ordering of these colors should appear very familiar, as they are identical to those commonly seen in natural phenomenon where white light is broken down into its constituent frequencies; the colors of the rainbow. Each color expresses a range of frequencies that fall within specific wavelengths of radiant information. The human eye detects these wavelengths and interprets them as the colors of the visible light spectrum. Physics texts illustrate the electromagnetic spectrum as a narrow band of radiant information just above "Infrared" and just below "Ultraviolet." Upon closer examination of this band, we see a spectrum identical to the colors viewed in the rainbow as light separates through a prism, or the human energy system.

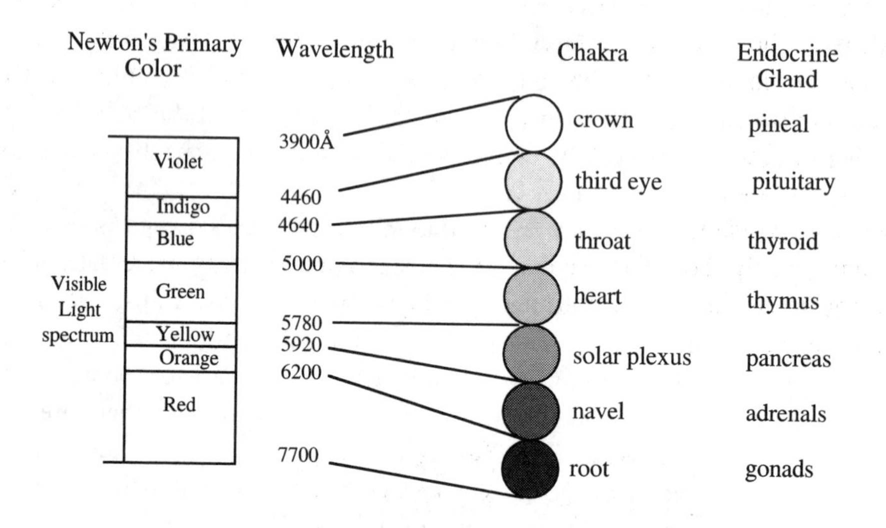

Figure 43 *The electromagnetic spectrum showing the "visible light" band expanded. Note the correlation between this spectrum and that of the human chakra system in Figure 42.*

The chakras are specifically designed to act as one level of tuned and resonant antenna, intercepting specific wavelengths of naturally occurring radiant information and stepping that information down to manageable frequencies for the systems of the body to utilize. Additional more refined tuning occurs at the molecular level, as genetic receptors called *codon triplets,* or simply codons (Chapter 2, "The *Seed of Life*: A Living Model of Knowledge"), receive information at an even greater level of differentiation. The spin rate of the chakra is part of this tuning, with the higher chakra centers rotating much faster than their lower counterparts.

For example, the Base or first chakra indicated as red, is resonating to vibrations associated with wavelengths occurring within a specific range. Light values for the Root chakra, then, are associated with Angstrom values between 6,200 and 7,700; a healthy Root chakra must maintain a minimum rate of rotation to achieve these values. These are the optimal frequencies that the endocrine glands, in this case the gonads, will respond to in their respective energy cycle.

A direct relationship exists between each individual chakra and the specific ranges of energy within our creation matrix. Chakras are our interface points, the energetic organs linking various aspects of our body to their nonphysical counterparts within the grids. Through this resonant relationship, we are offered access to the entire range of creation as an infinite matrix of information.

Through your chakra interface you are a part of all that you see. For example, the seventh chakra associated with the pineal gland, interfaces with an infinite grid of radiant energy-information-light, holographic expressions of itself as a pineal gland. It is through this interface that the reality of the human experience becomes apparent; our bodies may seem as much more than the independent and autonomous beings that we perceive. From this perspective, experience is a continuation of energy seeking expression through multiple zones of expression. The actual size of the chakra is relative to the individual, as shown in Leonardo da Vinci's drawing of human proportion (Fig. 44). The distance from the wrist to the tip of the longest finger is a function of the Fibonacci series of numbers and is referred to as a Fibonacci, or *Phi* Ratio. This particular ratio indicates the radius; one half of the diameter of the chakra cone. The distance that the etheric organs extend away from the physical body is also a relative distance equal to the radius or one *Phi* Ratio of a healthy vital being. The cone very quickly narrows to a small point directly above the associated endocrine organ. The scanners referred to earlier are able to detect and verify this point while displaying it through a video, audio and paper tape readout.

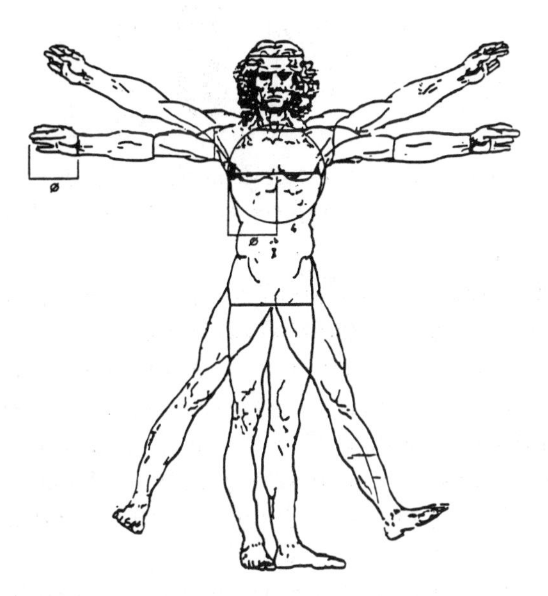

Figure 44 *Leonardo da Vinci's illustration of the human body indicating symmetry, proportion and ratios of the Fibonacci sequence. The distance from the wrist to the tip of the longest finger is equal to half of the chakra diameter.*

In aging or illness, the chakras have collapsed into the body and slowed their rate of rotation. The ancients knew of this, as evidenced in the Tibetan prescription of a series of movements or "rites" designed to maintain the vitality and spin-rate of the chakra. Recommended as a daily meditation, these movements are not aerobic in nature, though they may be performed quickly enough to become aerobic, and are used in the Tibetan monasteries as a mode of prayer, meditation and physical maintenance. In her recent book entitled, *The Ageless Body,*[1] Chris Griscom details these rites and their use, as well as additional movements, that have traditionally remained confidential within the monasteries.

Our knowledge may be considered as twofold in nature: that of the conscious experience acknowledged through the living senses on a daily basis, and that of the knowing that is received on a cellular level. These are unconscious signals integrated into our bodies and experienced "automatically" as an ongoing process. An illustration of the energetic hierarchy may help to understand this process that affects each individual every fraction of every second of each day, throughout the life experience. The radiant energy of creation rains incessantly upon our world and our bodies; wave packets of information coded as geometric patterns of subtle energy carried upon waves of broad spectrum radiant information: light! Each second of every day we are asked to process this information, determining what is useful and what is not. For the codes that are useful we are further asked to find the place within our mind-body-spirit complex where the patterns may reside as a fit of balanced information. The square pegs find the square holes and the round pegs find the round holes. This is not an insignificant task!

All chakra points are *exposed* to all wavelengths of information simultaneously. The key here is that not all wavelengths of information are *meaningful* to all chakras. Each chakra, through its spin rate, is *tuned* to vibrate within a specific range of frequency. These frequencies determine which portions of the light are meaningful. The chakra and associated grids respond only to the information to which they are tuned, allowing the remaining information to pass. It is through our resonant receptors of genetic code that we, individually and as species, are able to tune the chakras and access the living information of light as it passes through our bodies. All of the information of creation is always available. The information may not be meaningful, however, until we are able to bring it in through the tuning mechanisms.

Many individuals have unconsciously learned to assimilate information, in the following manner:

1) The information is received through the hierarchy of subtle bodies as the matrix becomes "thicker" toward the physical portion of the matrix.
2) The system, overwhelmed with the sheer volume of information, begins to temporarily store the data within "buffers" for the body to assimilate at a later time. Often, this time is somewhere within a 16 hour time frame.
3) The buffers will continue to store the information until they reach saturation; they can hold no more. At this point the buffers are vibrating at a specific tone that moves into resonance with the sleep centers of the brain, releasing serotonin (a compound counteracting the alerting effects of other chemical processes) inducing sleep within that individual.
4) The sleep process allows the body the time required to empty the information accumulated within the buffers and assimilate the energy into the matrix of the body—to the places where the life experience dictates that the information should fit. The fit will change throughout

the life experience as new experiences allow the reevaluation of previous experiences and a reorganization of how they fit into the system.

As our bodies remember to vibrate differently, resonating to higher expressions of information, a significant by-product is that we require less sleep. You may be experiencing this in your life at the present time. Collectively and individually, we have already remembered to vibrate more rapidly than we may have, even two years ago. Initially, an increase in cellular vibration may be accompanied by the *perceived* need for *more sleep*. This is especially true if deeply rooted emotions have been awakened. The need for additional sleep will dissipate as the body settles into the new patterns of vibration and assimilation of information.

Awakening our bodies to greater possibilities of vibration may be viewed as a two-edged gift; though emotions have become dislodged, forgotten and sometimes painful memories may surface. The higher vibration of experience will allow for these memories to be resolved quickly. In the resolution may be found the balance and healing. The increase in vibration allows the chakras to rotate at a faster rate and process the information in "real time" as the events are occurring. The net effect of this is that our bodies store less information and demand less sleep time to process.

The Science of Mudra
Access to Creation Through Body Circuits

Layers of circuitry connect the primary vortex points of our chakras to various portions of our bodies, both *internal organs* and "contact points" located upon the *surface of the skin*. In oriental medicine, these pathways of current are referred to as meridians or ley lines. It is along the meridian lines that precise contact points are identified and used in the ancient sciences of acupressure and acupuncture. Intersection points may be stimulated or short circuited, promoting or preventing impulses from reaching specific nerve centers. There is an entire science based upon the knowledge of techniques used to access the chakra vortex points through the meridians. In the Eastern texts one aspect of this science is developed through the connecting of body circuits in various combinations, to produce specific voltages within the body, and the associated frequencies. The study of these techniques is referred to as the science of *mudra*.

Located at the fleshy tip of each finger is an electrical contact point providing direct access into a chakra zone through one or more meridians. Each fingertip is associated with one primary chakra, although it may have secondary access to other chakras. Sequence is critical in accessing the appropriate chakra center. Following the use of the first four fingers, the finger circuits are reused to access higher chakras. The thumb serves as a ground and is touched to the tips of each of the fingertips to accomplish the desired mudra. For example, the mudra to access the root chakra is that of thumb and index or first finger. The second chakra is accessed as the second finger and thumb and so on. Upon completion of the fourth finger and thumb mudra, the first finger and thumb are used again to access the next higher, or fifth chakra. The fingertip and chakra associations are as follows:

Finger Designation	***Chakra***
Index or First Finger	Root or First Chakra
Second Finger	Second Chakra
Third Finger	Third Chakra
Fourth Finger	Fourth Chakra
In this order only, the first finger now accesses the fifth chakra	
First Finger	Fifth Chakra
Second Finger	Sixth Chakra
Third Finger	Seventh Chakra
Fourth Finger	Eighth Chakra

Table 7 *Finger/Mudra/Chakra Relationships.*

Recent meditative techniques offered in the West, as well as many Eastern techniques, use mudras in their meditations. As you have the opportunity to observe and experience these techniques, look closely at the mudra-body circuit-chakra-grid that they are accessing. Often it has been the first or second chakra only. It is from these grids that the frequencies of energy were being drawn and focused within the "lens" of the physical body, to be directed by intent.

The instruction of prayer often includes the placing of the hands in a special position to communicate with God, the Creator or to Creation itself. The mudra of prayer is that of the hands facing one another in front of the body, fingers extended with the tips and palms touching one another. The use of this mudra is very common throughout the world, and is taught as the preferred form of prayer within the Christian traditions of the West. What has been often overlooked, however, is the reason for the mudra. Each of the fingertips, through touching those of the opposite hand, completes the circuits and provides access to the associated chakra zones and grids *simultaneously.* The net effect is enhanced through the contact of the large chakras located within the center of each hand, the palm chakras. It is through these access points, the fingertips and palm chakras, that energy-information-light is directed to other parts of the body or to other individuals. For many, these techniques are almost instinctual, as they mirror a knowing from another aspect of their being. Through the science of mudra, prayer may become an intentional process of acknowledging a singularity with creation. Within the context of that oneness, you may intercede upon your own behalf, rather than asking for intervention from a power that you believe is beyond you.

Consciousness as Information

Figure 45 is a generalized schematic showing the path of information as it is stepped down through the matrix of creation toward the physical-nonphysical interface, continuing into the body. The source of information is any "wrinkle" in the matrix producing energy-information-light as broad spectrum and radiant information. The source may be a planet, a star or sun, the center of our own Milky Way Galaxy: that which the ancients referred to as the Great Central Sun. Propagated into denser portions of creation through tuned resonance, the information passes through process bodies, subtle blueprints of our form expressing as specific zones of the spherical matrix.

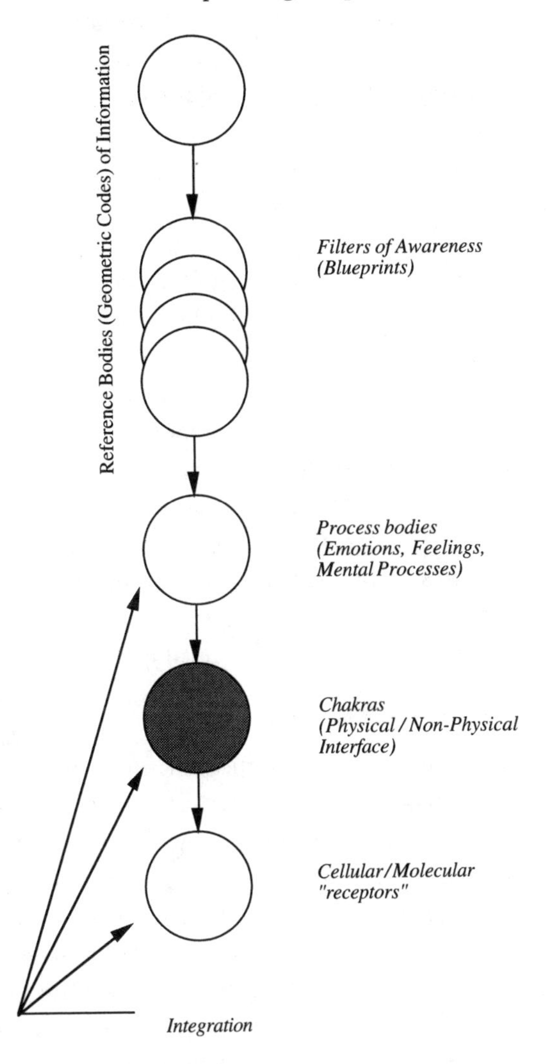

Figure 45 *Generalized model of the information path as it is propagated into denser portions of the creation matrix toward our bodies.*

These zones become the repositories for information relating to "mind" and "emotion," as mental and emotional bodies. It is within these zones of the matrix that we seek to discern between the emotions of experience and the experience itself. Perhaps one of the greatest lessons to be learned from this cycle of consciousness is that we are not our experience.

You are complete and whole regardless of your *perceived* success, *perceived* failures and shortcomings, "broken" relationships and "shattered" emotions. Each of these perceptions is a pattern of experience that provides indicators of growth within our lives. However they are not "us." Our soul-essence-life force is free and independent of these patterns until we bind them together through our perceptions.

From the process bodies, the patterns of information continue their journey through the matrix, radiating as encoded patterns of light. At this point, the patterns begin to express in the only manner that they "know" within the constraints of the third dimensional world; as the Platonic Solids. One of two distinct, yet related "light processors," begin to pick off the codes that they are tuned to, propagating them into the physical structure through specialized programs designed to recognize these codes.

The first of the processors, our chakra system, comes to life in receipt of codes tuned to their program. For example, the root chakra will be stimulated through the codes of light within the visible spectrum of 7,700 to 6,200 Angstroms; the heart chakra will respond best to 5,780 to 5,000 Angstroms. The codes intercepted by the chakra are then fed into the primary system regulating our body's functions; the endocrine system. It is the endocrine glands that comprise what has become the most fundamental, though perhaps the least understood, system of our bodies. Regulating chemical, pH, hormonal balance and body temperature, our mineral-based endocrine glands may be viewed as the first solid physical interface of our physical and nonphysical environment.

The second of the processors is that of the gene-cell complex. Within each cell of our bodies lie what may be thought of as *micro chakras,* those of the deoxyribose nucleic acid, or DNA. The relatively long and intertwined form of the double helix molecule actually contains the equivalent of many small, independent yet related chakras, that reside along the axis of the molecule. In the terminology of molecular biologists, these chakra points are expressed as sugars bonding to one of four possible base structures designated as "A," "C," "G" or "U." The sequence of these bases along each strand of the DNA molecule determines the makeup of the amino acids that are essential to life (Table 8). The bases are referred to as *triplet codons* or simply "codons." Though there are 64 unique combinations possible, with two exceptions, each combination fails to produce unique compounds.

Historically, the 64 possible combinations have produced only 20 unique compounds; our 20 amino acids. This is possible at present due to a phenomenon that researchers refer to as a redundancy in the code. In other words, multiple combinations of base pairs result in the same amino acid. Interestingly, however, it is the third member of the codon triplet that appears to be less significant, as if it has not fully contributed to the ordering. For example, consider rows three through eight of column one in Table 8. The amino acid *Leucine* is represented by the code UUA. It is also represented as UUG, CUU, CUC, CUA and CUG. Why? Why don't these six combinations of C, O, H, and N produce six unique amino acids? Why do 64 possibilities of the code produce only 20 unique compounds? Why is it that some of the amino acids appear to be more "generalized" than others in their makeup? The answer to these, and similar questions may be found in the environment within which the codes are expressing rather than within the codes themselves.

Chapter 2 reminds us that matter expresses uniquely in response to given parameters; the "environment" within which it is existing. The human genetic code, and that of all life, has historically expressed its specificity within the constraints of relatively dense magnetic fields of Earth and the relatively low frequency of ~8.0 cycles

	Col. 1	Amino	Col. 2	Amino	Col. 3	Amino	Col. 4	Amino
U	UUU	PHE	UCU	SER	UAU	TYR	UGU	CYS
U	UUC	PHE	UCC	SER	UAC	TYR	UGC	CYS
U	***UUA***	***LEU***	UCA	SER	UAA	BLANK	UGA	BLANK
U	***UUG***	***LEU***	UCG	SER	UAG	BLANK	UGG	TRP
C	***CUU***	***LEU***	CCU	PRO	CAU	HIS	CGU	ARG
C	***CUC***	***LEU***	CCC	PRO	CAC	HIS	CGC	ARG
C	***CUA***	***LEU***	CCA	PRO	CAA	GLN	CGA	ARG
C	***CUG***	***LEU***	CCG	PRO	CAG	GLN	CGG	ARG
A	AUU	ILE	ACU	THR	AAU	ASN	AGU	SER
A	AUC	ILE	ACC	THR	AAC	ASN	AGC	SER
A	AUA	ILE	ACA	THR	AAA	LYS	AGA	ARG
A	AUG	MET	ACG	THR	AAG	LYS	AGG	ARG
G	GUU	VAL	GCU	ALA	GAU	ASP	GGU	GLY
G	GUC	VAL	GCC	ALA	GAC	ASP	GGC	GLY
G	GUA	VAL	GCA	ALA	GAA	GLU	GGA	GLY
G	GUG	VAL	GCG	ALA	GAG	GLU	GGG	GLY

Table 8 *The Genetic Code for carbon-based life.*
(Adapted from *The Molecular Biology of the Gene,* James D. Watson.[2])

per second. The genes of the average person contain 46 chromosomes, with the codon matrix actually "coding" for roughly one third of the possibilities available, the 20 codons are turned to "On." The use of the remaining 44 codons has been a great source of mystery.

Recent studies indicate that three of the "unused" codons may be "start and stop" signals for the molecule. What of the remaining 41 codons? Could it be that these remaining codons are activated in response to shifts in the "environmental" factors of Earth's magnetics and frequency? As the parameters of The Shift become more pronounced through lower magnetics and higher base resonant frequency, the third member of each codon triplet has an unprecedented opportunity to express completely. Shift parameters, coupled with an individual's *willingness* to access more complex information, are allowing new chemical bonds to form, resulting in new forms of amino acids. Researchers have reported these genetic shifts in the open literature as *spontaneous mutations.* Greater base resonant frequencies allow variations of the existing amino acids; new "antennae" through which to move into full resonance with creation. As each of our 64 genetic possibilities is enabled, the code for the new amino complexes may appear uniquely as follows:

	Col. 1	Amino	Col. 2	Amino	Col. 3	Amino	Col. 4	Amino
U	UUU	PHE1	UCU	SER1	UAU	TYR1	UGU	CYS1
U	UUC	PHE2	UCC	SER2	UAC	TYR2	UGC	CYS2
U	***UUA***	***LEU1***	UCA	SER3	UAA	BLANK	UGA	BLANK
U	***UUG***	***LEU2***	UCG	SER4	UAG	BLANK	UGG	TRP
C	***CUU***	***LEU3***	CCU	PRO1	CAU	HIS1	CGU	ARG1
C	***CUC***	***LEU4***	CCC	PRO2	CAC	HIS2	CGC	ARG2
C	***CUA***	***LEU5***	CCA	PRO3	CAA	GLN1	CGA	ARG3
C	***CUG***	***LEU6***	CCG	PRO4	CAG	GLN2	CGG	ARG4
A	AUU	ILE1	ACU	THR1	AAU	ASN1	AGU	SER1
A	AUC	ILE2	ACC	THR2	AAC	ASN2	AGC	SER2
A	AUA	ILE3	ACA	THR3	AAA	LYS1	AGA	ARG1
A	AUG	MET	ACG	THR4	AAG	LYS2	AGG	ARG2
G	GUU	VAL1	GCU	ALA1	GAU	ASP1	GGU	GLY1
G	GUC	VAL2	GCC	ALA2	GAC	ASP2	GGC	GLY2
G	GUA	VAL3	GCA	ALA3	GAA	GLU1	GGA	GLY3
G	GUG	VAL4	GCG	ALA4	GAG	GLU2	GGG	GLY4

Table 9 *Possible configuration for new genetic code allowing greater resonant access to the creation matrix.*

In Table 9, the codon matrix is still intact with 64 possibilities. The three-letter triplets, however, have lost their generality. In this hypothetical example, each code is producing an amino acid unlike any other within the matrix. As the parameters of The Shift move forward, especially the increase in Earth's resonant frequencies, the contribution of each component of the triplet assumes greater importance. The third member of the triplet is allowed to express fully and completely, yielding new forms of our amino acids!

Let us examine this possibility in greater detail. At first glance, Table 9 may appear identical to Table 8. In the illustration, a dark-line box around LEU (Leucine) is offered as an example to illustrate the concept of unique coding. Chemically the aminos are the same; structurally each is expressing a higher, more complete order. The same chemical combinations result in the same essential acids. Closer examination will reveal that the same codon combinations are producing unique forms of the aminos (LEU1, LEU2, LEU3...). Exposed to more complex and shorter wavelengths of information, *the new light*, the third member of each triplet in full expression of itself, is providing unique qualities to the existing forms of Leucine. If human life is defined, in fact, genetically by the type and arrangement of essential amino acids, and these are changing, then by our own definition, a new life form may be emerging from the human species; a form for which there is no name at present.

Our world "is" what it "is" because of the information that we receive and process. We define our world through the lens of that which we have access to. Many individuals have chosen an existence that may be known through perception alone. That perception is allowed through the 20 codon "antenna." For those individuals, their perceptions of life are limited to the view afforded through the 20 codons that are turned "On." This concept becomes important in relation to an individual's perception to The Shift, as the availability of "off" or "on" codons is largely determined by the vibrational parameters of feeling and emotion that the body is in resonance to.

Remembering to vibrate differently, unplugging from old grids of limitation by allowing resonance with new and empowering grids, allows the development of a new collective consciousness. This new awareness does not recognize the same limitations or restrictions of the previous grids. Individuals finding resonance with these new grids view themselves from a much different perspective. Their lives have new purpose and meaning coupled with a sense of oneness with creation, rather than an experience of separation. As the effects of The Shift continue to reverberate throughout our lives, one tangible piece of evidence that we may expect to see will come from the science of genetics, as individuals who have "remembered" exhibit new forms of the existing amino acids.

From the level of the cell-gene, information is used in the most efficient manner that the body knows of at a given point in time. The degree of success with which the information is redistributed into our mind-body-spirit matrix determines the rate at which our bodies signal the willingness to process new information. Codes that find a "fit" within the grid-matrix structure patterns of energy that allow us to feel that we have "learned" (Fig. 45).

Within our tuned, resonant and holographic matrix of experience, our ability to alter energy within any level produces an effect that is felt through the remainder of the system. In other words, as one individual "learns," to some degree all have "learned." We have always possessed mechanisms providing such an ability. Subtle as they may appear, *our feelings and emotions* are our primary tools to shift the energy of creation in general, and specifically, within your creations. Figure 46 will help to illustrate how and on what levels thought is most potent, in terms of the impact that its energy has upon your ability to regulate consciousness.

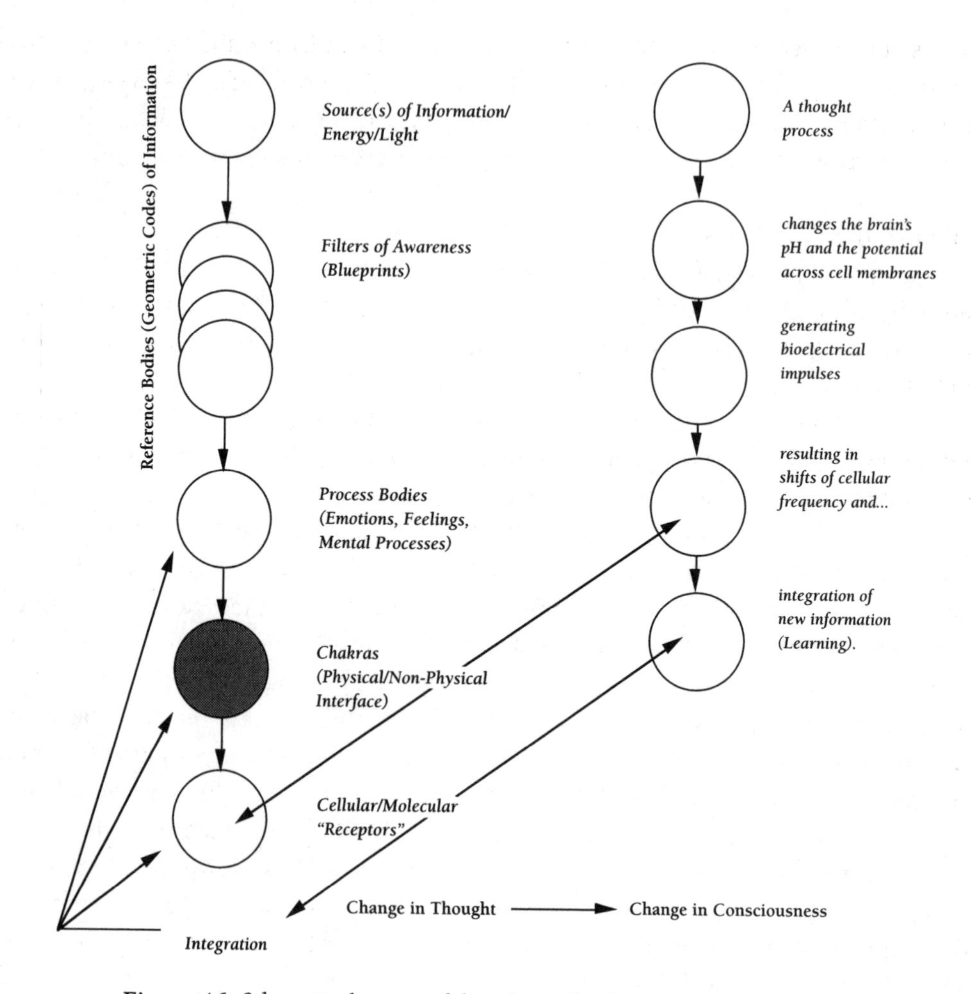

Figure 46 *Schematic diagram of the relationship between the matrix of consciousness and the process of thought and feeling.*

The left portion of the diagram is identical to that of Figure 45, the conceptual model of energy-information-light descending through the matrix into dense expression. To the right is a flow diagram illustrating the relationship and effects of thought and feeling upon the body.

The seed of a thought is initialized and begins attracting patterns of energy unto itself, vibrations that serve as patterns of constructive interference within the body, beginning in the brain. The shift in frequency produces a shift in the acid-alkaline balance (pH) within portions of the brain. This new balance is signaled to pertinent cells throughout the body as a shift in the electrical potential across the cell membranes. The net effect is an altered frequency of that cell, and the associated charge that it yields. The new frequency shifts the patterns of information and reorganizes them into new patterns with a new "fit." With regard to the information that is being reshuffled within the cell; it is said to have been *integrated*. This is the learning that occurs upon an unconscious level every fraction of each second every day. It is the change in thought that ends old cycles, old patterns and belief systems and allows new information to replace the old patterns.

This is our awakening. We awaken simply by *allowing* for other possibilities, other truths. This is not to imply that those possibilities are always condoned, agreed with or desired; they are simply allowed for. The key to our allowing is to "shift our awareness" to a viewpoint that redefines our interpretation of experiences through the eyes of fear and pain.

Cellular Knowledge

Our tools of cellular learning and the experiences of our lives are one in the same. It is through our quality of thought, feeling and emotion, *our belief systems,* that physical change within our bodies may be accomplished. Thought, feeling and emotion are the keys to altering the bioelectrical charge within the cell. Our experiences offer us feelings and emotions regarding that particular experience. To truly understand how each of those experiences affects the body on a cellular level, it is necessary to examine the relationship between thought and cell.

Each cell may be viewed as an endpoint, a dense physical expression of one end of the semiautonomous matrix of the human form. The cell receives a complement of energy-information that has been intercepted through the tuned chakra, then stepped down to a manageable frequency for that portion of the energy system. It is at this cellular level that our bodies may be viewed as truly electrical. Each cell within every organ, each of the one quadrillion (100,000,000,000,000) or so cells within the average human body, functions as a minute electrical circuit. Our cellular circuits exhibit properties of resistance and capacitance; they transmit, receive and store energy and information. Our cellular circuits are adjustable and tunable!

Figure 47 is a schematic illustration of a single human cell. It is composed primarily of fluids exhibiting varying amounts of electrical charge, or *potential.* These fluids are separated by thin porous membranes of tissue, allowing only certain fluids to pass at certain times. It is the motion of fluids between membranes, from one side to the other that allows the electrical potential to be realized, to become charged. It is in the distribution of fluids within the cells that the electrical nature of the cell becomes apparent. Each cell, then, becomes a bioelectrical generator functioning as a miniature electrical circuit. This charge fluctuates within each individual and is digitally measurable.

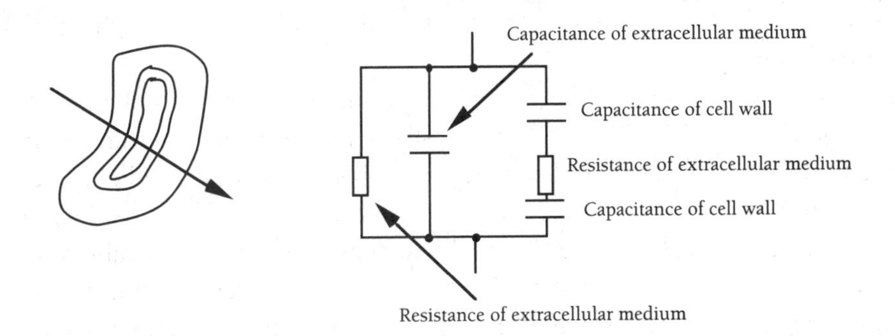

Figure 47 *On the left is a representation of a single human cell as a function of fluids separated by thin membranes of variable density tissue. On the right is a schematic illustration of the electrical equivalent of the same cell. Note the components of resistance and capacitance.* (Adapted from Schwan, 1953-1956.)

Each cell functions within a specific range of frequency, with each assemblage of cells tuned to a range or a zone of frequency. Any single organ operating within this range of frequency may be viewed as having a characteristic or *signature* frequency, a composite vibration of all cells within that organ. Our bodies, as a whole, radiate a characteristic or signature frequency, the composite of all cellular assemblages within. It is this signature frequency that will provide the key to addressing unwanted electromagnetic information, viruses, and so on, as an alternative modality of healing.

Within each cell of the human body, there exist balances on many levels; electrical, chemical, magnetic, and so on. One aspect of balance, that of the acid/base is of primary concern in this discussion. The numeric value indicating the degree of acid/alkaline balance between cellular solutions is known as the pH. pH is measured on a scale from 0 to 14, 0 being the greatest acidity and 14 indicating the greatest alkalinity. A reading of 7 is neutral, indicating a perfect balance between the two extremes. Thought processes, cellular frequencies, and the ability to regulate these frequencies, are functions of the brain's pH.

As we experience, there is an emotion or a feeling that is associated with our experience. One portion of that feeling is the actual electrical charge of the thought. A change in our thought produces a change in the pH of that particular portion of our brain. This change, in turn, alters the pH of the cellular fluids within certain portions of the body, increasing or decreasing the electrical potential of the fluids separated by the cell membrane. This dynamic relationship governs the rate at which the cell is vibrating; the cellular frequency. This fundamental principle, though often not realized, is used as therapies of "positive thought" and affirmation.

From this simple model, the value of meditation becomes apparent. As we embrace specific belief systems, the patterns of energy that mirror those beliefs may or may not serve us. Through meditation, the haze of those patterns is bypassed, even for the relatively short duration of the meditation itself, allowing the introduction of a new pattern—a new belief—into the system. As we remember that we have access to the subtle energetic blueprints of experience, we grasp the powerful realization we may change the patterns of life that no longer serve us. The very act of intentional change within us *is evolution*; conscious evolution. It is through the science of Compassion that a neutral charge may be generated (pH 7) regarding a situation or individual.

This neutrality does not necessarily condone, agree or sanction the experience witnessed. Rather it allows for a redefinition of what our experience has shown us. Through our redefinition of hate, anger, judgment and jealousy, as well as love, Compassion and forgiveness, we remember the ancient opportunity that allows us to move forward in life as empowered beings of intent.

SUMMARY

- Our experiences may be thought of as the accumulation and reorganization of energy-information-light seeking balance.

- Information is stepped down incrementally through the matrix of creation into a denser portion of the same matrix, the familiar form of our bodies.

- Thought and feeling directly impacts information within our matrix-body through regulating cellular frequency and the integration of information (learning and remembering).

- The effect of our cellular integration is echoed back through the matrix as balance, signaling the readiness to receive and integrate additional information.

- We have the ability to "choose" the quality of our thoughts and feelings. Inherent within our choice is the opportunity to create healthy and vital patterns of information within our bodies.

- Thought, feeling and emotion regulate the pH of our brains which, in turn, determines the chemical balance (voltage) and frequency of each cell within our bodies.

- Molecules, such as the DNA, respond to these altered frequencies as follows:
  - New frequencies are generated and sustained at molecular/cellular levels.
  - The chakra system resonates closer to its optimal frequencies.
  - The aura reflects all of these changes as its vital radiance.
  - New forms of the essential amino acids begin to develop as the codes are allowed to express fully.

CHAPTER FOUR

Crop Circles

Resonant Glyphs of Change

Speak to the Earth and it shall teach thee.

THE HOLY BIBLE, JOB 12:7

Within the holographic model all knowledge is eternally present, stored as resonant patterns of vibration and form. As *individual segments* of consciousness develop the ability to recognize this information, each element of *the whole* benefits to some degree. Herein lies the beauty of the holographic model. Many individuals whole and complete unto themselves, contribute by virtue of their experience as facets of an even larger whole. Following the model, as one remembers, to some extent all remember. As one heals, to some extent all may heal. Through the holographic experience, a relatively few number of individuals may create change through *becoming* the change. Though new possibilities of experience are always present, they sometimes lie beyond the collective frame of reference and may not be recognized until the experience of life provides a blueprint for the meaning.

For example, author and researcher Tom Kenyon relates an account of a meeting between Spanish explorers and the indigenous tribes that greeted him in Africa. The explorers were en route to the New World, sailing around the tip of Africa in massive ships driven by large canvas sails. The ships anchored a safe distance offshore and used long open boats powered by oars to bring a small crew ashore. Never having seen white men or their ships before, the natives asked the men how they arrived in that part of the world. The sailors pointed to the ships anchored offshore with the large white sails. Looking in the direction that the sailors had pointed, though they tried, the natives could not see the ships. It was the shaman of the tribe that tried something new. He discovered that in using his eyes differently, the image of the ships began to "appear" in his view. As he practiced this new way of seeing, others soon began to "see" the ships also. Within a short period of time, other members of the tribe could see the ships, *even those who were not there to meet the crew as they rowed ashore.* The tribe had learned a new way of seeing and in doing so had expanded the boundaries of their perceptions. This account is a beautiful example of collective resonance, often identified as the hundredth monkey principle. Once a single member of the whole achieves, to some degree everyone else benefits.

The purpose for relating the story is to illustrate the idea that the "information" of the ships was always there; it did not change. The particular pattern of information represented by the ships was not familiar to the tribe, however. Beyond their individual and collective frame of reference, they had nothing to compare the ships to. There was no reference pattern to make sense out of the array of wood planks, canvas sheets and metal plates. Though their eyes and brains may have detected the images, their minds averaged the information out as extraneous and the experience was limited.

I believe that the events in this story apply to the present, as well as the indigenous tribe 500 years ago. We have been similarly conditioned through our own culture and society within this lifetime. As we take information into our eyes and pass the signals to our brains, culturally, we are taught to "look" for familiar patterns of information. We seek something that makes sense in terms of previous experience. The practice of "seeing" what is actually present has been conditioned out of our experience through locking onto the accepted behavior patterns of a reality that is expected. We call this generalized experience *consensus reality.* Each time we seek to place an experience occurring beyond our agreed upon parameters into a category that says the experience never occurred, we encourage consensus reality.

In "looking" at patterns of information that we are conditioned not to "see" or accept, there is a good chance that those patterns will not be seen. In addition to the seeing, it is in the sometimes unconscious recognition of symbols that there develops an exchange of energy between the radiant patterns of the information outside of the body and the coded patterns stored within our body-mind-memory complex.

From Simple Circles to Complex Pictograms

Though first reported in the 1600s, a phenomenon reoccurred in 1975 and continues to the present, with an increasing degree of mystery each successive year as to both the meaning and origin of the enigma. Glyphs produced in open fields, primarily those of cereal grain crops of wheat, corn and rye are appearing with greater frequency, complexity and in greater numbers, with the passing of each successive crop season. Commonly referred to as "Crop Circles," the first glyphs to arrive were relatively simple, infrequent and almost immediately labeled as hoaxes. It soon became clear, however, that all of the markings could not be hoaxes. If for no other reason, the sheer number and complexity of the glyphs would make them very difficult to produce. The technical sophistication, undisturbed ground beneath the crops, altered crystalline structure of the plants and tremendous respect for the crops themselves point to a phenomenon that is not produced with any technology known to exist at present. Markings known as Crop Circles have appeared in nearly every country that grows a harvest of cereal grains; including Australia, Japan, Russia, Canada, and now the United States. The majority of the glyphs have appeared in England, however, within a small area dominated by the remnants of ancient temple sites such as Avebury and Stonehenge. Appearing infrequently at first, the glyphs increased in both number and complexity until the 1990 season when the term "pictograms" began to be used to describe the mysterious markings.

Table 10 shows the geometric increase of pictograms through 1990, with 1991 and 1992 producing slightly fewer in number with greater degrees of complexity. The evolution of the patterns, coupled with the rapid increase in numbers, has led many researchers to believe that the markings are telling us something; offering us information regarding the relationship between our planet and its inhabitants. In consideration of the marked increase, in both numbers and complexity, the message of the patterns appears to be urgent! What are these mysterious markings? Where do they come from? How are they formed? Perhaps most importantly, *what are they saying to us?*

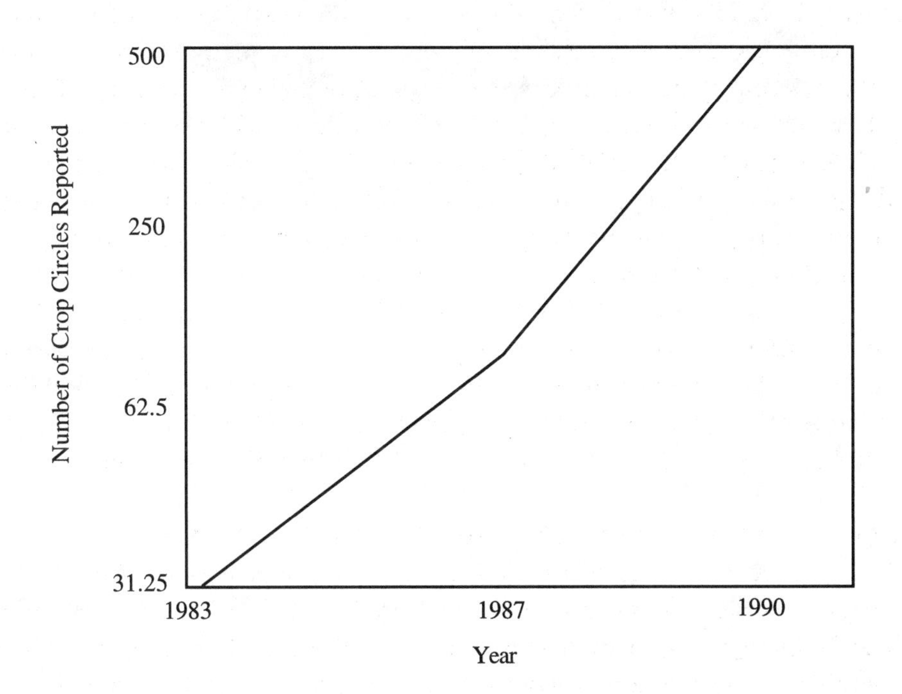

Table 10 *The nonlinear increase in the appearance of "Crop Circles" from 1983 through 1990.* (Data from Colin Andrews, *Undeniable Evidence,*[1] 1992. Graph by Sacred Spaces/Ancient Wisdom.)

The physical alteration of the crop patterns is an enigma unto itself. The glyphs are usually round, sometimes imperfect, with the stalks of the crop lying on their sides arranged in a spiral pattern. A tremendous respect is demonstrated for the life of the plants, as the stalk itself is seldom broken, continuing to grow and produce its fruit.

Often the soil beneath the crops is that of fragile clumps resulting from the heat and moisture patterns within the area. Upon investigation, the clumps are extremely brittle and crumble to the touch. During the formation of the glyphs, this fragile ground is not disturbed, the clumps are not touched, broken or rearranged. The spirals themselves are often interwoven, describing the patterns as alternating layers of clockwise and counterclockwise stalks (Fig. 48).

Figure 48 *Detail of patterns showing the clockwise and counterclockwise interleaving of stalks.* (Courtesy of Sharon Warren, ilyes, Ron Russell and the Center for Crop Circle Studies.)

Microscopic analysis, through the *Signanalysis Laboratories* in Stroud,[2] indicate that the actual crystalline structure of the plants within the circle is quite different from that of control samples taken from other sites or the same site outside of the glyph itself. The control samples indicate less order in the crystalline pattern of the cells, compared to a higher order of crystallinity expressed through the samples taken from within the glyphs.

THE FORMING OF THE GLYPHS

The markings themselves show great diversity from season to season and field to field. Historically, the early glyphs of each season have been primarily simple circles varying in size. Over the years, the circles began to appear more complex, with rings surrounding them; single rings, double and then triple rings. Each ring had formed concentrically around the central circle. With each legitimate Crop Circle, the delineation between the edge of the glyph and the remainder of the field is always very well defined. One of the mysteries of the phenomenon is the precision with which the forms are placed in the fields. Leading Crop Circle researcher Colin Andrews reports that the formations appear as if "cookie cutters" had been used to delineate the glyphs. Eyewitness accounts report no traffic at the time the glyphs are formed, no sign of footprints, vehicles, or human activity moving into or out of the crops. There are reports of some glyphs forming in a relatively short period of time, with individuals standing in or near the fields as they are formed. Occasionally reported is a sound or "noise" that precedes or accompanies the birthing of a formation, sometimes accompanied by a bright yellow light in the sky above the fields.[3]

The fundamental relationship that allows our physical world to appear as it does, is that of vibration (frequency) and form (geometry). This relationship may best be defined as

"Frequency yields geometry and geometry yields frequency," *a two-way relationship.*

Frequency <==> Geometry

This relationship may best be stated as: form is a function of vibration and vibration is a function of form.

Creation exists portrayed as energetic patterns of wave form, aligning themselves along predefined guidelines of expression. It has been demonstrated through the science of Cymatics[4] that repeatable and predictable patterns may be found in substance as a direct response to vibration. It is unlikely that all of the patterns forming in each of the fields are "hoaxed" as the technology is not acknowledged on Earth today that will produce patterns of the intricacy, complexity or sheer number that are seen in the fields.

Witnesses have, however, reported sounds associated with the formation of some glyphs. Sometimes the sounds are barely audible, appearing as sustained tones overlaying additional tones prior to or during the formation of the glyph in the crops. To my knowledge, no one has actually witnessed the stalks of the plants bending in response to any force, mechanical or otherwise to form the patterns, and they may not, for one reason. If the glyphs are, in fact, being formed through the knowledge of the vibration-form relationship, the entire phenomenon would occur so quickly that it is unlikely that anyone would observe the event unless they were already present within the field at the precise moment, in the precise location of the glyph. It is my belief that the frequency-geometry relationship is the technology responsible for the *"how"* in terms of the glyphs in the crops.

CLASSIFICATION: THE SEARCH FOR PATTERNS

In the 1978 crops, what appeared to be the first cross was formed consisting of four small circles arranged symmetrically around a single larger circle. The stalks were lying down in a counterclockwise direction with the "fruit" of the crop unharmed. By 1990, the glyphs had become so complex that they were no longer referred to as Crop Circles; rather the term "pictogram" was coined to describe the increasing degree of complexity being shown in the symbols.

By their very nature, the markings in the crops defy true classification at present. Any attempt to compartmentalize these mysterious, yet familiar, symbols is arbitrary at best. Even so, the logical aspect of human awareness appears to "need" some way to impose order onto the mystery of their formation. For the purposes of study, one system of classification is offered here, based purely upon pattern recognition. Though they may not be completely understood, there are groupings of patterns that appear to be related and very dissimilar to other patterns. Sometimes the differences appear in glyphs located within the same field. As the glyphs become more complex, new aspects of the patterns may cause them to fall into multiple categories. Guidelines for a possible classification system are as follows:

Type I

Unique, one of a kind patterns with obvious significance and provability.

1) The Mandelbrot Set.
2) The Barbary Castle formation.
3) The "Broken Serpent."

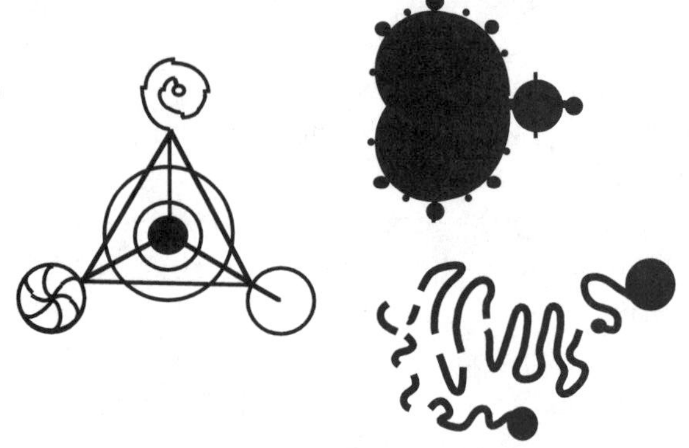

Figure 49 *Type I Patterns.*

Type II

Patterns with direct visual reference to the ancient science of Sacred Geometry.

1) The *Vesica Piscis* or the "womb" of creation formed by overlapping two perfect circles such that the outermost perimeter of one passes through the center point of another. The common area between the two circles is the "vesica."
2) A six-petaled star, the *Seed of Life* formation indicating counterrotating fields of electrical and magnetic information.

Figure 50 *Type II Patterns.*

Type III

Patterns consisting of two or more circles, indicating a relationship with one another, often connected.

1) Each of the circles may be nearly identical.
2) One of the circles may be differentiated through the use of concentric circles, arcs, shading or appendages.
3) The "connector" may be linear, "tube-like," or "arc-like" as an elongated *Vesica Piscis*.

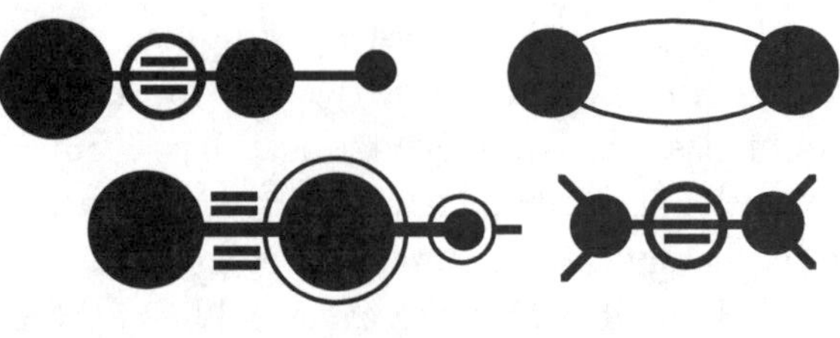

Figure 51 *Type III Patterns.*

Type IV
Stand alone, nonconnected, close proximity and "scattered" circles.

1) Some with one to three concentric rings around them.
2) Alternating directions of spiraling described within the layering of the crop within the glyph.

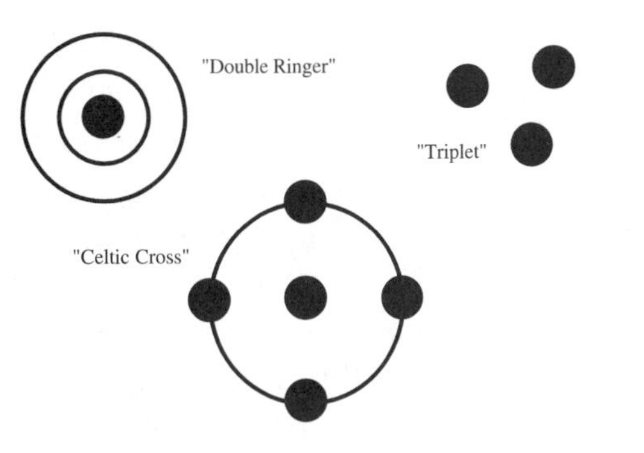

Figure 52 *Type IV Patterns.*

Type V
Stand alone, nonconnected, "Mandala" type glyphs.

1) The "charm bracelet (s)" I and II
2) The pentagonal "finale" for the 1993 season.

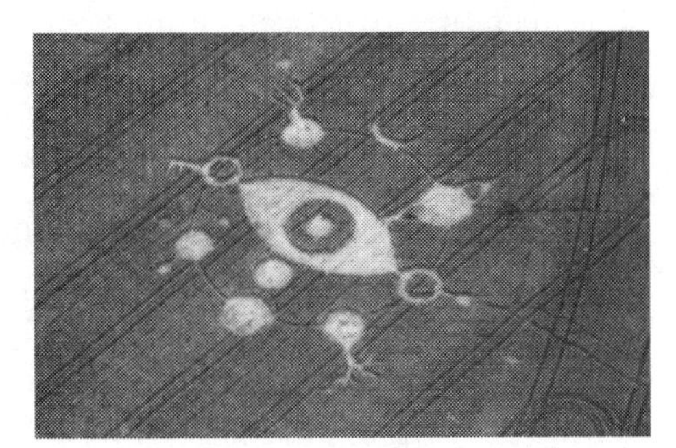

Figure 53 *Type V Patterns.*

Within the context of this simple classification scheme, the diversity of these formations becomes apparent. As the glyphs themselves become more numerous and complex, the message becomes clearer; it is a message of change. The beauty of these Earth inscriptions is that they are not limited to any one understanding or belief system. Special tools are not required to "know" the codes and work with them. Regardless of the walk of life, those that recognize the patterns, *know* them.

Of major significance in the Crop Circle enigma is the opportunity that now, for the first time in conscious human memory, everyone may participate in the phenomenon through direct experience. The glyphs are arriving upon Earth, intact, and may be viewed that way until they are altered or destroyed. The message is there for all to see, without interpretation, compartmentalization or a "higher authority" intervening on anyone's behalf to aid in the understanding. The glyphs are speaking to everyone, symbolically and directly.

Decoding the Glyphs: The Message

To understand the intent of the messages in our cereal crops, it first becomes necessary to recognize the language(s) of the offering. Initially, these fascinating symbols appear to many as just that; beautifully symmetrical images of precision, immense proportion and mystery. Some individuals have stated, "*We are not capable of under-*

standing the meaning" of the messages in the crops. Others, though they do not readily recognize the symbols, feel intuitively that there is a message, and the message comes as a warning to humanity to care for one another and the planet.

Upon examination of these "agri-symbols" it becomes apparent, purely through pattern comparison, that some of the glyphs are much more closely related than others. The symbols, artificial and intentional, are obviously conveying something in some language—but what? What is the language? How does the language work? More importantly, what is the message? Is it consistent throughout the phenomenon?

To add an additional component to the mystery of decoding the glyphs, in my studies I have detected at least five distinct, independent symbolic languages rather than a single language encoded within the glyphs over the years. These languages are living languages of resonance; meaningful patterns of information that we respond to, rather than the "vowel-consonant-symbol" languages used in the English and ancient alphabets. An example of the languages recognized, to date, are as follows:

- SACRED GEOMETRY
  Our language of Earth-Heart-Mind resonance

- GENETICS
  Symbolically and as resonant patterns of genetic code

- ELECTRICAL CIRCUITRY
  Direct symbols of the electrical circuit, some found additionally upon temple walls and ancient pottery markings

- MATHEMATICAL SYMBOLS
  Graphic representations of universal constants representing intelligence as we know it through our technology today

- SACRED SYMBOLS OF THE ELDERS OF THE EARTH
  Native American, Celtic and Egyptian

Possibly one of the most intriguing aspects of the crop codes is that as varied as they appear in their form, on a general level, all are conveying a very similar message, with varying degrees of clarity.

The message is this:

We are a part of all that we see in our world. Our world is undergoing a Shift that is echoing change throughout every aspect of our tetrahedral, carbon (human) and silica (earth) experience.

The Human-Earth component of our 12 planet solar circuit has reached a long prophesied, long awaited threshold of Entropy.* Earth and Human are changing

*Entropy may be defined as a measure of the capacity of a system to undergo change.

rapidly. That change is happening now. The shifts resulting from the change are occurring within the *genetic patterns* of our cells, and mirrored as physical, measurable shifts within the *electrical and magnetic circuitry* of Earth, as predicted through the ancient texts and prophecies. Specialized languages of symbol, languages that we have developed, toward the understanding of our world and ourselves, the very "codes" that are seen as Crop Circles, are reiterating universal messages of life; messages mirrored in the wisdom of ancient prophecy. The codes appear within the physical portion of our Earth-matrix-circuit in response to languages that are embedded within our human-matrix-circuit. These messages are echoes of our own technology, both internal and external, ancient and modern. The messages comes as no surprise, nor do the languages through which they are offered.

Human populations are unarguably diverse. In their level of awareness, knowledge and understanding, as well as application of their understanding toward their daily lives there are tremendous variations. This diversity is inherent in the beauty of the holographic model; each individual is whole and complete within his or her own knowing and, at the same time, part of an even greater whole, also complete within its knowing. For the "circle makers" to tailor such a universal and profound message as that seen in the Crop Circles, to any one belief, organization, belief system or technology would be to defeat the very purpose of the offering. Understanding the "language" is one component of the messages in the glyphs. The alphabet is nearly meaningless, however, without some frame of reference within which to read the message.

The second aspect of decoding the language becomes the *context* within which the message is offered. Evidence indicates that Earth is experiencing the precursors of dramatic change, collectively referred to as "The Shift." Each aspect of our lives plays a key role of a component of the process. It is to these specific aspects of our tuned-resonant-circuited being that the messages of the crops are directed. Now, the glyphs in the crops are physically validating externally, that which we have suspected and felt, possibly without the vocabulary to express. Our world is experiencing the close of a great cycle of conscious human experience! Examples of the type and degree of specificity encoded within the symbols follows:

- Details of Earth's shift patterns of capacitance, maps of "grounding" points within the circuit of our local solar system/Earth/human matrix.
- Specifics of shifting genetic code as points of greater specification in codon pairing.
- The fulfillment of Native American prophecy that marks the completion of the "Fifth World" of consciousness.
- Specific replication points of genetic material within the human DNA molecule.

Because the patterns of the Crop Circles interact with each individual directly through resonant codes, the interpretation and meaning varies accordingly. Through the unique lens of individualized experience, the emotional filters of those experiences, conditioning and culture, determine how the glyphs are viewed.

For example, in the prophecies of the Native American traditions, it is stated that humankind will know when the time of "purification" has begun through certain signs. The Hopi speak of the coming of a "blue star" and the dancing of Kachinas "in the plaza." Other indigenous people tell of a time when the moon will be seen, both on Earth and in the heavens. It is said that tears came into the eyes of one clan of Native American elders upon seeing the glyph illustrated in Figure 54.

Figure 54 *"The moon will be seen, both on Earth, and in the heavens."* (Quotation from indigenous prophecies.) *Glyph of the crescent moon.* (Courtesy of Sharon Warren, ilyes, Ron Russell and the Center for Crop Circle Studies.)

Others, who do not identify with the tribal prophecies, saw the same glyph as the fulfillment of a different prophecy; the Biblical reference:

> *And (it shall come to pass in the last days), I will show wonders in heavens above, and signs in the Earth beneath.*
>
> THE HOLY *BIBLE*, THE ACTS OF THE APOSTLES 2:19

Though the cultures differ, to each group, the sign of a crescent moon appearing upon the Earth has a significant, powerful and similar meaning.

A Mathematical Constant: Order and Chaos Converge

Possibly one of the most astounding glyphs to appear in recent years is that of the Mandelbrot Set. Appearing approximately ten miles south of Cambridge, England on August 12, 1991 this formation was the first that could actually be scientifically substantiated by researchers.

The study of chaos theory defines a point expressing the boundary between order and nonorder. It may be said that at this point order breaks down into chaos, *or conversely*, chaos gives way to a higher order (chaos being a degree of order within itself). The mathematics to describe this point, known as the Mandelbrot Set, are complex. Until recently, the mathematics were all that we had to reference this energetic boundary. With the development of high-speed computer capabilities, massive computer memory and fractal output, it is now possible to view the Mandelbrot equations graphically. The image portraying the point at which order and chaos converge is shown in Figure 55. Figure 56 is an aerial view of the glyph that appeared outside of Cambridge, England in 1991; that of the Mandelbrot Set.

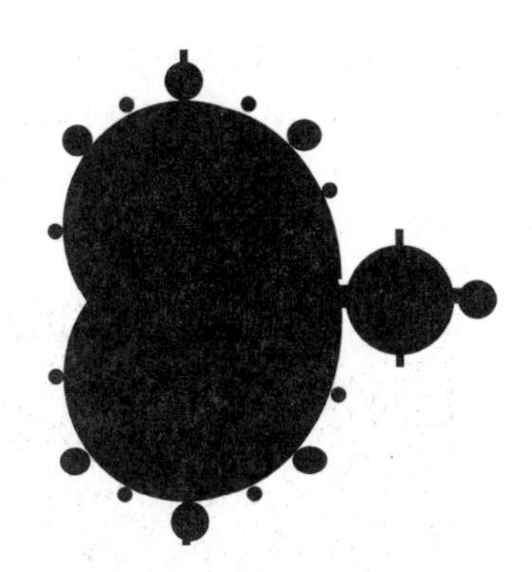

Figure 55 *Computer representation of the Mandelbrot Set.*

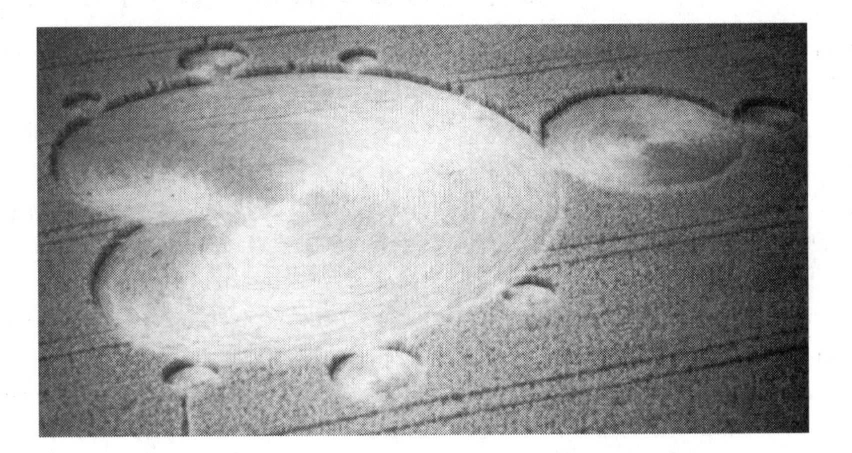

Figure 56 *The Mandelbrot crop glyph that appeared outside of Cambridge, England, in August of 1991.* (Courtesy of Sharon Warren, ilyes, Ron Russell and the Center for Crop Circle Studies.)

As Colin Andrews states in his video, *Undeniable Evidence*,[5] this is a particularly significant image because it portrays graphically and describes mathematically the precise point that the indigenous people of the world, as well as the ancient texts, indicate that Earth is at in this moment of time; the point in which chaos and order converge. Stated another way, we are experiencing the chaos of old belief systems and obsolete paradigms, giving way to a higher order characterized as peace, cooperation and harmony. In any case, it is the Mandelbrot Set that will probably be viewed as a turning point in the study of Crop Circles, as this pictogram was the first to be recognized universally as the graphic portrayal of a relevant and significant mathematical constant.

Genetic Shifts: Fine Tuning the Code

In August of 1991, what I believe to be one of the most significant glyphs appearing to date, arrived in a field near Froxfield, England. To the best of my knowledge, this unique pictogram has never been repeated, or even approximated, and was destroyed less than one week following its initial "discovery." Commonly referred to as the "Serpent," and also the "Brain" (Fig. 57 left), in and of itself, I believe that this glyph offers both a key, and a tangible map, of human genetic response to The Shift.

The modern science of genetic research has adopted a series of conventions to express, and map, relative positions of the basic building blocks of life; amino acids, proteins and the site of individual genes on a particular chromosome. Improved research techniques have yielded graphic models, *circular models*, of the genes, the amino acid sequences and the DNA molecule itself. A number of glyphs in recent years appear to be providing information regarding the fundamental nature of human genetics, including "weak links" in the structure and where we may expect to see tangible evidence of genetic shifts in response to the changing planetary parameters (magnetics and frequency). Again, the messages are offered to us in our own languages of symbol, mirroring our chosen path of technology.

When I first saw the "Serpent," I felt intuitively that this glyph contained more than a message—perhaps a clue—as to anomalies that we are witnessing in our collective gene pool, anomalies such as the mystery of codon specificity. From the 64 unique choices available, for example, why do our genetic programs "code" for only 20 of the choices? What about the other 44? Table 8 is an alphabetic map of the genetic code for carbon-based life. Within each physical cell of the body resides the relatively long and intertwined form of the double helix molecule, deoxyribose nucleic acid, or DNA. Situated along the axis of the molecule are many sites where bonding may occur that result in the forming of specific amino acids. This bonding occurs as sugars detect a chemical "mate" with one of four possible bases designated as "A," "C," "G" or "U." The resulting compounds are referred to as triplet codons, or simply codons.

There appears to be redundancy within the code, as multiple combinations of these codons that result in the same amino acid. Interestingly, however, the third member of the triplet codon appears to be "less significant" at present, as if it has not fully contributed in the ordering of the codon choice. It is within the matching of these codons that the mystery occurs: Of the 64 unique combinations available, each with the opportunity to produce a unique structure, only 20 compounds result. These are seen as our 20 essential amino acids.

This unique glyph, the "Broken Serpent," may hold the key to these and many other questions regarding the human genetic response to The Shift. Presented as a series of irregular loops, the Broken Serpent is composed of ten segments, each of varying length. Though some segments approximate the lengths of others, there are no identical segments. Table 11 indicates the segment lengths, with I.D. numbers corresponding to the break points on the glyph itself in Figure 57 left. The column labeled "% of Formation" is a breakdown of each segment as a percentage of the overall length of the glyph. Analysis of these lengths at present does not produce any recognizable sequence.

Break I. D	Scaled Segment Length	% of Formation	% of Circular
1	4.6	8.34	2.19
2	19.8	35.9	11.7
3	3.3	5.98	13.26
4	5.3	9.61	15.8
5	4.7	8.5	18.05
6	1.2	2.17	18.63
7	1.9	3.44	19.53
8	2.7	4.	20.83
9	2.6	4.71	22.08

Break I. D	DNA Site Mapped
1	Large rRNA
2	Cytochrome Oxidase II
3	ATPase subunit 6
4	URF 3
5	URF 4
6	URF 4
7	URF 5
8	URF 5
9	URF 5/6 (boundary)

Table 11 *Analysis of the nine breaks for the "Broken Serpent."*

Figure 57 right is an accepted map of human DNA, showing sites and ranges of the essential amino acids, as well as other significant regions on the molecule. Converting the relative percentages of each segment length from the "Broken Serpent" to a percentage of the circular mapped DNA (third column of Table 11), it becomes possible to "map" the break sites of the glyph onto the schematic of the molecule. Figure 57 right indicates the "Hit" sites as circled numbers.

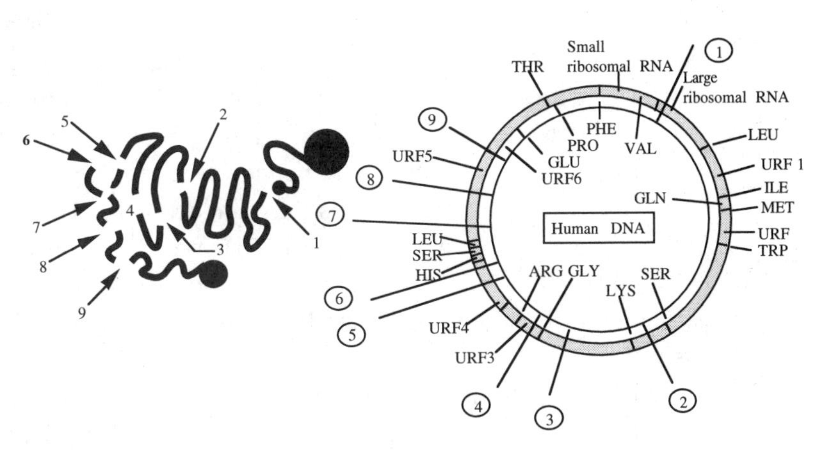

Figure 57
Left: *Schematic survey of the "Serpent."* (Adapted from Beth Davis.[6])

Right: *Circular map of human DNA molecule.* (Adapted from Norman Rothwell.[7])

Of the nine break sites, one is an area known as ribosomal RNA (rRNA), one plots into an area known as Cytochrome Oxidase II, one is in ATPase subunit 6, and the remaining plot onto URF (unassigned reading frames) 3, 4 and 5. While all of these locations are significant, the six URF locations are particularly interesting. URF sites mark zones within the DNA molecule that appear to be "unused," at this time, in the coding of the amino acids. If, for some reason, these sites are shut down (breaks), coding will not occur. If new proteins, resulting from new amino acids are to occur,

(Man is in the process of changing to forms that are not of this world...

THE EMERALD TABLETS OF THOTH,[8] TABLET 8)

these sites provide the "staging area" for the building of these structures.

Chapter 3 discussed the significance of existing forms of amino acids "learning" new forms of expression. To understand why human genetic codes are not expressing to their fullest capacity, it is necessary to look at the environment within which they are expressing. The clues are found, not within the codes themselves, rather, in the environment within which the codes are expressing. It is well documented that matter expresses uniquely in response to a given parameter; the environment within which it is existing. As the energy of the environment shifts, the crystalline expression of the matter must shift accordingly to accommodate the new environment.

Our genetic code, and that of life that we recognize, is composed of unique patterns of energy that have expressed their specificity within the constraints of dense planetary magnetic fields, and the relatively low frequency of approximately 8 cycles per second. As the parameters of The Shift become more pronounced (lower magnetics and higher base resonant frequency), the third member of each codon triplet now has the opportunity to respond as the complete expression of its code. The Shift, coupled with the individual's willingness to access more complex information, will allow new chemical bonds to form, resulting in new forms of amino acids.

Now, from something as seemingly unrelated as a glyph in the crops, for the first time we may have the genetic "map" showing us where we may expect the new forms of the codes to appear. Time will certainly tell.

Electrical and Magnetic Shifts: Fine-Tuning the Circuit

In the terminology of 20th century electronics, the term *capacitor* describes a device which stores an electrical charge that has accumulated to a certain point, releasing the charge, as a pathway becomes available to do so. Physically, it consists of three portions; two conductive surfaces separated by a nonconductor, or *dielectric*, surface. From an electrical perspective, Earth may be viewed essentially as a large spherical capacitor. The surface of the planet acts as one conductor that is negatively charged, portions of the atmosphere serve as a positively charged zone and the thin layer of air close to the ground provides the insulation or dielectric effect. Researchers have known this for years and utilize these principles in the transmission of energy and some communications devices.

The language of electronics uses special codes and symbols to indicate capacitance in blueprints, circuit designs and technical journals. Interestingly, the symbols are not really new, as they appear in ancient circuits displayed on temple walls in Egypt, as well as the mysterious designs and paintings of Native Americans on their pottery and weavings. Figure 58 is a sampling of the symbols, ancient and present, used to indicate electrical capacitance.

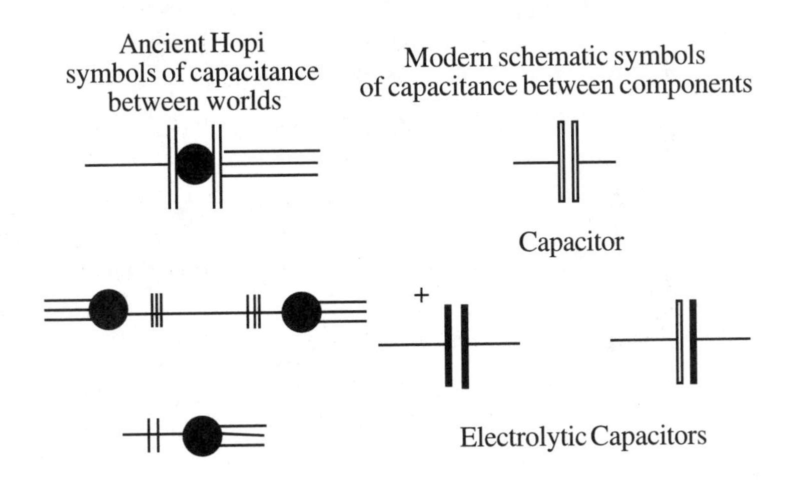

Figure 58 *Examples of ancient and modern symbols used to indicate electrical capacitance.* (Adapted from *Designs on Prehistoric Hopi Pottery,* Jesse Walter Fewkes,[9] *Understanding Electronics,* R.H. Warring.[10])

Beyond the dissimilarities of first glance, many of the crop glyphs, primarily of the Type III variety, have a common denominator; simply the patterns that are present. There appear to be two or more spheres, connected in some manner through a linear tie. Though the two spheres are very similar, one of them will exhibit a distinguishing mark that will set it apart from the first. If the convention of electronic symbology holds true for the glyphs, a number of these "connections" appear to be capacitive in nature. In Figure 59, note the descriptive symbols aligned with the linear connections between the spheres.

In each of these pictograms, there is an obvious relationship being described between the two spheres. One aspect of the description may be found in the sets of parallel bars aligned with the axis of the spheres. If the convention of 20th century electronic notation holds true, our own language of symbol, the glyphs may be describing a capacitive relationship between the spheres. In Figure 59 A, for example, the capacitive sphere "above" is exhibiting some characteristic(s) that delineates it from the sphere below, although both are connected through the vertical linear "tie." Figures 59 B and C also appear to be indicating a linear relationship between spheres, without the emphasis of difference between the two. Type III glyphs appear to be telling us at least two kinds of information:

1. The "spheres," whatever they represent, are connected, and experience some linear relationship.
2. The connection may be "electrical," possibly capacitive and resonant in nature.

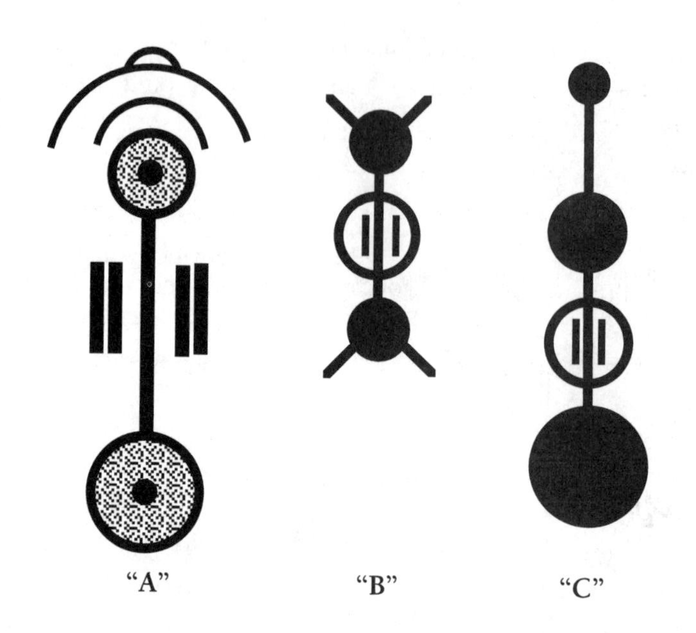

Figure 59 *Type III glyphs of connected spheres exhibiting linear "capacitive" connections. Each of these pictograms appeared prior to 1991.*

Again, this should not be surprising, as the languages of the glyphs continue to mirror our language, our symbols and our terminology across a multitude of disciplines. To those not familiar with these languages, the glyphs appear to be beyond our frame of reference (natives unable to see the sails). They may relate better to other symbols with a greater relevance to their life experience.

In terms of these particular glyphs, we may assume that because the glyphs are appearing upon the Earth, at least one of the spheres represents the reference upon which it has been placed; that of Earth. With respect to Earth and the process that is unfolding at this time, we know these things to be true:

- Earth is capacitive in nature.
- As a capacitor, it will build and store an electrical charge until it finds a pathway to release that charge.
- Historically, Earth has vibrated at approximately 7.8 Hz. That number appears to be increasing, with a predicted threshold of 13 Hz.
- The build and discharge cycle will bring Earth to another experience of itself.
- As a part of the experience, Earth is resonantly connected to additional celestial bodies, tuned primarily through our own sun.

As Earth continues to move into The Shift, it may be viewed electrically as accumulating a greater potential to be stored until a certain point is reached. That point of discharge will be the completion of the process, and the beginning of the event, the 180 degree reversal of the magnetic poles accompanying an increase of Earth's fundamental vibration to 13 Hz. Driven by the increased density of photons as the planet moves through the cyclic Photon Belt,[11] the shift of these two parameters will elevate the Earth sphere into resonance with a new zone of information; the capacitive discharge of the sphere at 13 cycles per second.

In the holographic model of localized creation, the capacitive circuitry of Earth may be viewed as a system, whole and complete unto itself. This system, in its wholeness, also constitutes one electrical module in an even larger system, that of our 12-component solar system. In viewing our solar system as a tuned and resonant circuit, each component is attempting to compensate for external conditions, such as the Photon Belt, in an attempt to remain tuned to the reference node of our sun. Assuming that the planetary model of this circuit is valid, each planet, acting as one finely tuned component in the larger solar circuit, must achieve and maintain a capacitive value that will allow it to remain in a stable relationship within the circuit. If the conventions of 20th century symbology are valid in the crop glyphs, this series appears to be showing us the present and future relationship of Earth, another energetic body, and possibly with itself. Earth appears to be connected to at least one other planetary body, through a path of resonance and capacitance; possibly indicating that a capacitive discharge will allow a specific, and resonant, point of balance: The Shift.

SACRED GEOMETRY: THE TEMPLATES OF CHANGE

For over ten thousand years, the ancient and mystic science of Sacred Geometry has preserved and provided the knowledge of the relationship between our bodies and the crystalline forms of creation; patterns that are constant, predictable and repeatable. These patterns remind us of the codes of creation, the framework within which matter forms (crystallizes) into that which we recognize today as our world. Crossing the lines of the broad categories defined earlier in this section, patterns representing geometric codes appear throughout many of the crop glyphs.

90 miles from the city of Luxor, Egypt, on the west bank of the Nile river, lies a temple complex that is unlike any temple found at any other location upon Earth. Predating the ancient remains of the temple overlying it by several thousand years, the architectural style of this site is unique, its origins uncertain. Upon the load-bearing walls of this temple are a series of patterns appearing as varied groupings of interlocking circles. Shared between the two circles is an area of overlap, the sacred geometric spaces referred to in the ancient texts as the *Vesica Piscis*. The vesica may be thought of as a zone of "tension" between two adjacent experiences. In the sacred sciences, this tension is the expression of separate and diverse aspects of creation seeking balance through a common experience; precisely the philosophical point that many individuals feel that we are approaching collectively; The Shift. This glyph, indicating the desire for balance, appeared as two interlocking circles for the first time in the 1993 growing season (Fig. 60 right).

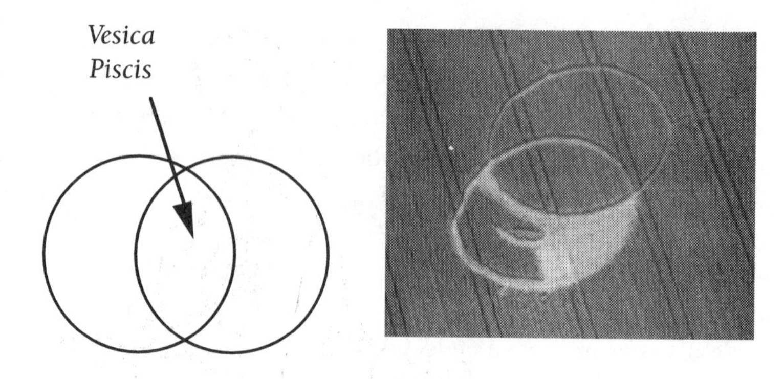

Figure 60
Left: *Schematic diagram of the* Vesica Piscis, *the "womb" formed through the union of two circles.*

Right: Vesica Piscis *glyph in English crop, 1993.*

On the temple walls, each complete pattern, or flower, is composed of 19 interlocking circles (multiple *Vesica Piscis*) with the entire pattern enclosed within two nested, concentric rings. Tangent to the outermost circles, the rings are referenced through the science of Sacred Geometry as the *Zona Pelucida,* a Latin term for the inner and outer walls of the human ovum. The Zona Pelucida may be considered as a metaphor for the potential of all possibility existing as masculine and feminine experience. Throughout the ancient teachings, the overall pattern is known and referred to as that of the *Flower of Life* (Fig. 28).

Located within the heart (center) of the *Flower of Life* is found the fundamental pattern of the *Seed of Life* (Fig. 61 left). Composed of six "petals," careful examination will reveal that each petal is the union of two opposing curves. Ancient mystery schools expressed the unified fields that "drive" creation through the science of opposing forces; the sacred union of opposites. It is through the regulation of this union that the forces of creation, in general, and life, specifically, are mastered.

In our field-grid-point-matrix experience of life for example, the forces represented are those of the pulsed electrical and magnetic fields. Thus each petal of the *Seed of Life* is the union of electrical and magnetic information: electromagnetic energy as "crystallized" patterns of energy-information-light. The *Seed of Life* is the "core" of the *Flower of Life*, the beginning of change. Without the seed there can be no flower. Without experience, there can be no change. Figure 61 right, is a glyph showing the derivation of the six-petaled flower, as it appeared in England late in the summer of 1994. Viewed from the perspective of Sacred Geometry and The Shift, the *Seed of Life* mirrors our position within the collective experience of change.

Figure 61
Left: *Schematic illustration of the Sacred Geometric form, the* Seed of Life.

Right: *The* Seed of Life *glyph as it appeared in the fall of 1994.* (Photograph by Steve Patterson, The Studio, Winchester, Hants.)

Figure 62 *Schematic illustration of the Barbary Castle glyph, as it appeared in July of 1991.*

The Barbary Castle formation, possibly the most recognized of the formations, is undeniably geometric in nature (Fig. 62). Central to the figure is the four-sided polygon known in sacred science as the *Tetrahedron*. The least complex of the five Platonic Solids, the Tetrahedron is one of the most fundamental patterns that energy uses to align itself into crystalline, "solid" matter. The mineral kingdom is one obvious expression of these patterns. The mineral of silica (common quartz) makes up over 90% of Earth's composition. Atoms of silica and oxygen bond in specific packing arrangements to crystallize, expressing as strings, chains and rings of tetrahedrons (Fig. 63).

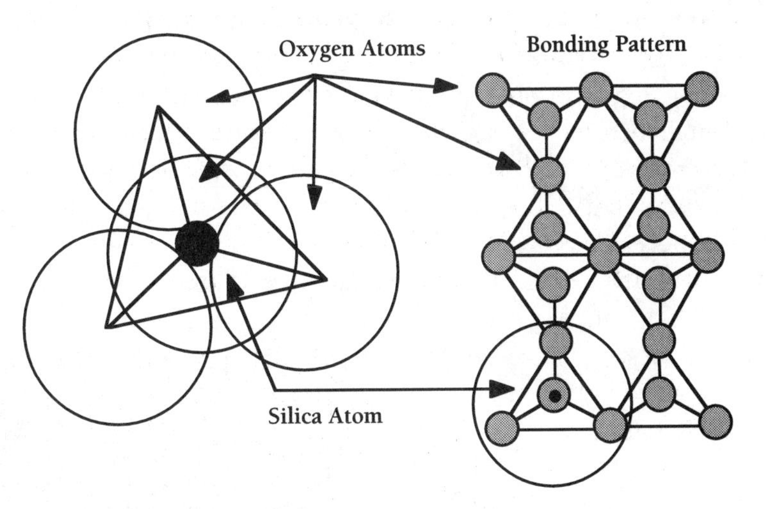

Figure 63 *Tetrahedral patterns of bonding to form quartz.*

Perhaps not so apparent is the role that the tetrahedron plays in the human body, and the human energy system. The ancient science of Sacred Geometry reminds us that we *live* the geometry of creation. From the instant of conception, our cells divide within the womb following precise, predefined patterns of form; living geometry! This is the "Sacred" aspect of Sacred Geometry; our very form mirrors the process of creation. Figure 64 is a schematic diagram of the geometric arrangement of the human embryo, after the first two cell divisions: the four-sided four-cornered form of the tetrahedron. The radiant field of *prana* (life force) from this packing arrangement is a key structure to the human link with the matrix of creation.

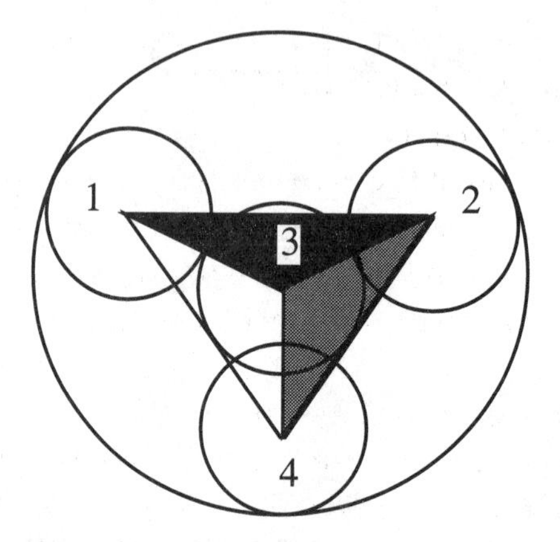

Figure 64 *Schematic illustration of the geometry expressed by the first four cells of the human embryo, following two mitotic cell divisions within the womb. A series of straight lines connecting the centers of each new cell result in the Platonic Form of the Tetrahedron.*

The physical aspects of our bodies are composed primarily of carbon compounds, the elements of carbon and oxygen account for over 99% of our bodies. These elements combine as the bonding form of the tetrahedron. It is not surprising, then, that the crop glyph of the tetrahedron has become the most acknowledged and recognized pattern of this 18-year-old phenomenon. What is the message of this extraordinary glyph and what does it mean to us?

The message of Barbary Castle is directed to those forms of energy-matter that lie in resonance with the tetrahedron; the carbon-based human form and the silica based form of Earth. In and of itself, the message of this figure appears out of place; a tetrahedron with circles at each corner of the form overlying a series of concentric circles. Within the context of the previous three languages, as well as the positioning of Earth in the completion of this cycle, the tetrahedral glyph of Barbary Castle begins to look much different.

Through the science of Sacred Geometry, parameters allowing Earth and the three dimensional experience are noted, by convention, as the universally accepted "language" of rotational and counterrotational fields. This notation results from the literal interpretation of the human *Mer-Ka-Ba* or Light Body Vehicle of Space—Time: Two identical fields of information, occupying precisely the same location at precisely the same time—rotating in opposite directions. The rotation to the right, clockwise, is the rotation of the magnetic portion of the electromagnetic fields surrounding the body. This is frequently expressed as an "arc" to the right) and in the ancient sciences is associated with the feminine aspect of creation. To the left, the electrical, or masculine portion (of the field is expressed again as an arc, in the counterclockwise direction. Combined, the two opposing arcs express a complete electromagnetic signal. This information, in union, provides the "media" through which creation expresses itself within the context of our time-space-dimensional world. The identical concept is conveyed through the language of Sacred Geometry as the union of polarities, the masculine and feminine field forming the area known 0 in the Sacred Sciences as the *Vesica Piscis*. With these principles in mind, it becomes apparent why the six-petaled pattern of the *Seed of Life* is so meaningful. The seed represents the pattern of wholeness expressed as the union of electromagnetic information, our experience, described through the cyclic language of Sacred Geometry. Reexamining the *Seed of Life* from this perspective (Fig. 61), what is indicated are six complete "petals" of the flower, each expressing union of electrical and magnetic information.

This very glyph was presented to Earth in the Winchester, Hampshire area in the fall of 1991, indicating the full circle, Alpha and Omega (beginning and the end) of the electromagnetic experience; one half of the petals to the left and the remaining half to the right.

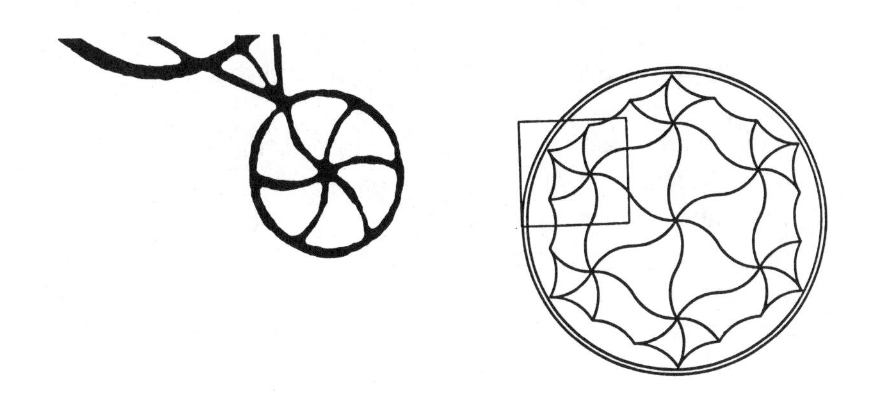

Figure 65 *The illustration to the right is the "magnetic" portion of the electromagnetic spectrum indicated by the six-petaled* Seed of Life. *To the left we see the magnetic field of the tetrahedron as indicated in the Barbary Castle glyph.* (Computer graphic courtesy Bruce Rawles.)

If we remove either component from the flower, the remaining pattern is exactly one half of the field; the electrical portion or the magnetic portion. The magnetic portion, to the right, would appear as Figure 65 left. Serving as an "anchor," stabilizing one quarter of our electromagnetic experience through the tetrahedral fields of experience, Barbary Castle provides an indicator of the component of our experience that is undergoing change. The counterclockwise arcs have been removed, mirroring the digital reality of our experience; declining fields of magnetics. Matter and life, tuned to the tetrahedron, are moving toward a Zero Point Magnetic experience.

To the left of the magnetic spiral in Figure 65, at the second corner of the tetrahedron is a curious spiral, indicated in a clockwise direction. In the ancient traditions of many of the indigenous peoples of Earth, the symbol for "light" is often shown as a continuous, though offset, arrow or line. Ancient sciences predicted, and modern science now suspects, that light does not travel in a straight line. Within the grid/vortex framework of creation, geometric evidence suggests that "light" moves as a series of curves following a very special form of connected curves; the logarithmic spiral. The figure to the right of the magnetic spiral, then, may be interpreted as that of a curved path of light, curving in the direction that is most in harmony with the spirals of light emanating from the radiant source at the center of our own Milky Way. It is toward this pattern of light that Earth is moving, in an attempt to resonantly match the reference signals generated as we pass through the "new light" of the electromagnetic zone known as the *Photon Belt.*[11]

The remaining two spheres, one completely dark, the remaining one completely light, are the universal convention accepted for polarity; positive (+) and negative (-). These four parameters, in and of themselves, are largely responsible for the experience of Earth, the "crystallization" of matter into the forms that we recognize as our third dimensional reality, as well as the "crystallization" of life force within the womb. The common denominator is the template of the tetrahedron. With these parameters in mind, a reexamination of the Barbary Castle formation begins to provide a new interpretation of this meaningful, yet mysterious glyph. The message may read as follows:

The fundamental parameters that allow Earth to be the Earth that we know and expect, are shifting; magnetics, light and the polarity of the planet (North [+] and South [-]). The Shift is addressing those aspects of experience that are in resonance, "tuned," to the grid and matrix patterns of a specific geometry; the tetrahedron. As tetrahedral bonding patterns of silica (Earth is over 98% silica) are experiencing The Shift, each aspect within our matrix retunes to resonantly align with new reference patterns. The tetrahedral bonding patterns of our bodies (the human form is over 98% carbon) are experiencing The Shift as each aspect within our matrix attempts to resonantly align with what, historically, has been our reference pattern; the resonant fields of Earth!

This field is responding to new "stimulus" by shifting, and in doing so, our lives and our world are changing, reaching for expressions of themselves as greater and higher balance. This is precisely the message of the ancient texts and the prophecies of the indigenous peoples of the world.

The message of Barbary Castle is a very ancient message presented in the symbolic language of our own technology in a format that cannot be ignored. Within this particular glyph, however, the message becomes more specific.

Whether the source and significance of the pictograms and circles are ever acknowledged is irrelevant at this point. If, for some reason, no individual ever witnessed another glyph in the crops, the whole would continue to be affected through the properties of resonance. If the holographic model is valid, as one has "seen," to some degree, all have "seen." The symbols have been seen, recognized and understood. The collective whole has benefited and will continue to do so as a result. Realistically, the patterns will probably continue to appear each growing season until the close of the cycle. In an unobtrusive manner, the glyphs are awakening patterns of choice within each individual; patterns that some are not even aware of. The pictograms are a gift to be acknowledged and shared by all as tangible evidence of at least one additional level of intelligence within our experience that is communicating with us, as well as, to us:

The glyphs speak to us symbolically, and directly!

SUMMARY

- Though the origin, mechanism and meaning of the symbols in the cereal crops of Earth may be debated, three factors are certain and cannot be denied:
  1) The number of the glyphs increased geometrically until 1991, apparently peaked during that year, with fewer, though more complex patterns appearing since then.
  2) The patterns have "evolved" from what appear to be simple circles to tremendously complex pictograms, also at a geometric rate.
  3) The glyphs are exact, intentional and *not* naturally occurring.

- If the glyphs are, in fact, a message—the urgency of the message appears to be greater now than in years past.

- As isolated phenomenon, the appearance each growing season of mysterious patterns in cereal grain crops may be viewed as interesting and curious. However, viewed in the context of Earth's positioning within the 200,000 year cycle, the enigma of crop glyphs takes on a much different meaning.

- Crop glyphs are offering to our world a manifestation that may not be hidden and can no longer be ignored.

- The glyphs in the crops, purely by virtue of their existence, are radiating patterns of information into our awareness matrix, creating change, passively!!

- Two fundamental categories of "codes" offered are:
  1. Resonant patterns, recognized on a cellular-genetic level, as the geometric codes within the structure of the cells themselves.
  2. Resonant patterns, recognized on a conscious level, as the symbolic languages of our science, inducing change within the cells through altering "beliefs."

- The languages recognized to date by the author, are those of:
  1) Sacred Geometry (Language of the Heart)
  2) The genetic code (both symbolically and as resonant patterns of genetic code)
  3) The symbols of the electrical circuit
  4) Graphic representations of mathematical constants
  5) Symbolic language of the elders of Earth

- All of the languages appear to be conveying the same message on a high level, with additional glyphs providing details on specific topics.

The message: The Human/Earth "component" of the 12 planet solar "circuit" has reached a long prophesied, long awaited threshold of entropy. Earth and the human form are changing rapidly. The change is happening now.

CHAPTER FIVE

Message of Zero Point

Ancient Seed of Opportunity

And ye shall never endure the pains of death: but when I shall come in my glory ye shall be changed in the twinkling of an eye from mortality to immortality; and then ye shall be blessed in the kingdom of my father.

THE BOOK OF MORMON,[1] 3 NEPHI 28:8

Discussions of our Zero Point experience are incomplete without mention of one individual whose lifetime, by design, demonstrated to us the possibility of our greatest life potential. Through the living memory of one man we were reminded what it means to fully enable our genetic codes and to do so through our choice of life conduct. One individual remembered, and through his memory, we are offered a point of reference, a standard reminding of our greatest possibilities.

Nearly 2,000 years ago a single event marked the beginning of a lifetime that would forever change the course of human history and the way in which individuals view themselves and their world. The birth was that of a man, an extraordinary man, through the womb of a woman who had not given birth previously and was virgin at the time. From the onset, the event was marked with unusual circumstances. The arrival of a new "star" in the heavens was heralded as the long awaited sign of a great prophet. Fear of this sign by those in power resulted in the deaths of thousands of Hebrew infants in an effort to nullify the prophecy of this "savior." Each of these was an extraordinary event by any measure. The unusual life of the man known as Jesus, the son of Mary and Joseph, is well documented through the early years of his life. Young Jesus was recognized by the elders of the Synagogue as being very illuminated in the interpretation of the scriptures and their application in daily life. To date, the best records of the early years of Jesus are those found in the Biblical texts. The subject of controversy for hundreds of years, the discovery in 1947 and subsequent translation of the Dead Sea Scrolls has validated the Bible in all but a few of the Chapters known as the Old Testament. It is from the Biblical references that the life and teachings of Jesus are preserved and interpreted today.

Though distorted to some degree by virtue of many translations, the events of Jesus' life are documented in the open literature until the age of 12, at which point nearly all Biblical references disappear for approximately 18 years. The references reappear at age 30, as he resumes a very public life of teaching and healing. The 18 "lost years" of Jesus, though unaccounted for in the present form of the Biblical texts, may have been recorded in earlier versions prior to their reorganization by Constantine's Council of Nice in 325 A.D.[2] It was during this time that many significant Biblical references were deleted and the remaining books reordered into present forms of the modern Bible. There are additional indications that these deleted books, stored in the Great Library of Alexandria, were lost in the burning of that structure in 389 A.D. References to the message and life of Jesus were so significant, however, that they were recorded and stored within other great libraries, preserving valuable insights into the life, teachings, travels and intent of this remarkable man.

Within the Jemez monasteries in the city of Leh, near the Kashmir region, are records of a great prophet that came from the holy lands. This prophet was noted as the first to master, as well as study, the teachings of Buddha, Krishna and Rama. These records, the originals of which remain today in Lhasa, Tibet, detail the life of this prophet. His name was Ehisa, and he journeyed throughout India, China, Tibet, Persia and Egypt before returning to his home in Israel at the age of 30, proclaiming that "my father and I are now one." According to the Tibetan records, the travels and studies lasted for 18 years.

Possibly the most tangible evidence of Jesus that exists today are the remains of an image "burned" into a burial shroud. This woven relic is said to have wrapped the body of Christ Jesus following his Crucifixion. Stored in Turin, Italy, the Shroud of Turin (Fig. 66) is made of linen measuring 14 feet 3 inches by 3 feet 5 inches and has been the subject of debate, skepticism and mystery for centuries in regard to its authenticity.

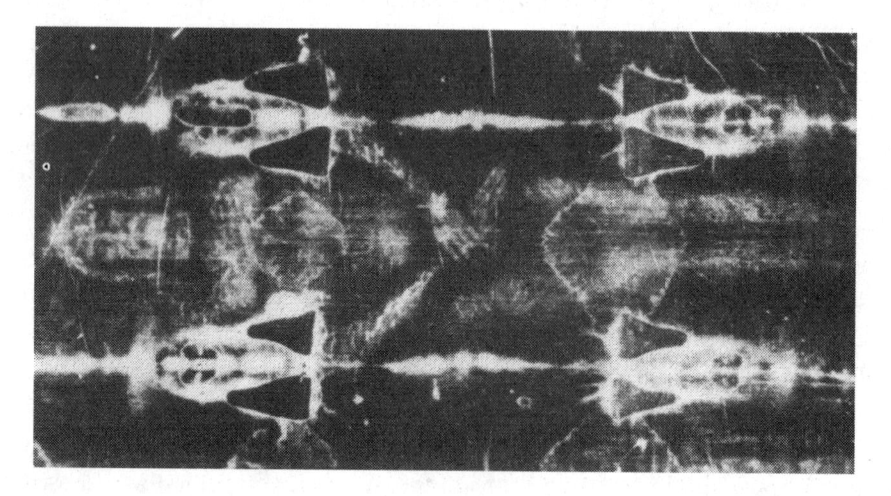

Figure 66 *The Shroud of Turin with the imprint of a crucified body and injuries matching those of Christ Jesus. The image has been "flash burned" into the cloth, as a photographic negative, from the inside of the linen.*

In October of 1978, a team of researchers at the Los Alamos National Laboratories, led by Ray Rodgers,[3] demonstrated that the image upon the Shroud is the equivalent of a photographic negative, and is not painted or dyed into the fabric. Inexplicably, the image is literally *burned* into the cloth as the result of a high-intensity flash of electromagnetic radiation originating *from within the Shroud.* The details of the Shroud match, item for item, the Biblical description of the body in terms of the type and location of wounds to the wrists, feet, abdomen and head. It is the opinion of Ray Rodgers, and his team of researchers, that the Shroud is, in fact, an authentic image of the body of Jesus preserved through an unexplained process of intense biochemical radiation approximately 2,000 years ago.

Other spiritual texts record the visit of a "white prophet" during a time correlating with the lost years and a message very similar to that which Jesus instilled into his followers following his reappearance. A number of Native American traditions, for example, trace the roots of their belief systems directly to the traditions of the ancient Essenes. In both North and South America, oral traditions relate stories of a bearded Essene prophet who came to them from the East with the message of Compassion and reverence for all life. Through the traditions of the Hopi, for example, the message may best be exemplified through their reminder:

> *Give love to all things, mountains, trees and rocks for the spirit is one though the kachinas are many.*
>
> ADAPTED FROM *MEDITATIONS WITH THE HOPI*[4]

In many of these traditions, the prophet promised to return toward the completion of this great cycle of experience with a message from the "Father." The Hopi prophecy relates the story of "Bahana," the white brother who promised to return to those that he taught during his lifetime. He is regarded as the "purifier" and will spare from destruction those who have maintained the Hopi way of peace.

The foundation of the *Book of Mormon* is that Jesus walked the Americas and anchored his teachings firmly with others as well, in the West as well as his home land. On his 30th birthday, the man known as Jesus of Nazareth reappears in the records during his initiation by John the Baptist. It is written that John, foreseeing the coming of Jesus as the Messiah, had spoken to his followers saying:

> *I baptize you now with water for repentance, but he who is coming after me is mightier than I, whose sandals I am not worthy to carry; he will baptize you with the holy spirit.*
>
> *THE HOLY BIBLE*, MATTHEW 3:11

As Jesus approached the river where John was performing baptismals, John immediately recognized him and spoke, saying *"It is I who needs to be baptized by you, and do you come to me?"* Jesus replied, *"Let it be so, now."*

It is during the years following his baptism that Jesus became known for his teachings and probably best for his "miracles" of healing and manifestation. In the face of his truth, those fearing his message were powerless to silence him—short of taking his life. In attempting to silence his teachings, Jesus of Nazareth was crucified under the direction of a Roman official, Pontius Pilate, in the year 33 A.D. The irony of this action may be seen in the intent. The Crucifixion, in an effort to silence, actually anchored the teachings and life of this man even more firmly into the memory of humankind. In rising (Resurrecting), three days following his "execution," Jesus demonstrated a rebirth, an eternal life reaching beyond his apparent death and the fear that prompted it. Through his Resurrection, he also modeled a process that each individual has the opportunity to experience within this lifetime, that of Earth's Resurrection: The Shift of the Ages.

Historically, the word "Christ" has been a reference to Christ Jesus of Nazareth, though other highly evolved reference beings, *other Christs*, preceded Jesus, some by thousands of years. One difference setting Jesus of Nazareth apart from the others was that earlier Christs appeared as *fully realized beings* specific to a race or tribe. Buddha, Akenaton, Shiva, Gogyeng Sowuhti, each planted seeds in human consciousness, paving the way, preparing for a time in their future; a time of great change in the life expression of man. With that change, each alluded to a great messenger, the Christ for all of humankind, regardless of race, geography or tribe. This messenger would be the Universal Being of Reference, our Universal Christ. The message that the Universal Christ would carry was to be one of remembrance, human destiny and purpose. The term "Christ," as used throughout this text, is a reference to the Christ Jesus of Nazareth. Intended as a message of hope and opportunity, the possibilities demonstrated through the brief lifetime of Christ Jesus transcend religious doctrine and teachings based in separateness, favoritism, rules and dogma established later through distortions of his original teachings. This is a message for all of humankind, present or past, reminding us of our possibilities through the sacredness of all life.

CHRIST'S GIFT TO EARTH

The gift that Jesus offered to Earth was in the form of a message to all of humanity, a message delivered and anchored by virtue of his life experience. While each of the preceding Christs had a similar message and could have served as a Universal Christ, it was Jesus who became the Universal Christ by design, *choosing* to emerge into this world in a manner that we could relate to; a birth through the womb. By doing so he demonstrated that he was birthed and lived within the same parameters as those around him, with no apparent tools of divinity other than his faith. Jesus lived the life of his time, associated with the people of his time, learned a skilled trade and the use of economics, ate the food and slept in the homes of those around him. Recent evidence suggests that he married and fathered at least one child, a daughter. Through his life he demonstrated that he possessed nothing more than any other individual, except the knowledge and faith of his nature and potential. Through the use of two very powerful tools, available to each man and woman, he was able to effect change within himself, the world around him, and ultimately transcend the perceived limitations of that world. The tools were the gifts of "Choice" and "Free Will."

Etched into our ancient memory is the belief that Jesus died for the "sins" of humanity. It is taught that through some mysterious process, known only to a few, the death of this man upon the cross served as a *sacrifice* for all of humanity and that through his sacrifice, all of humanity now has the opportunity to follow in his footsteps as perfected beings. Herein lies the foundation of the mystery and the key to understanding the relevance of his offering to Earth's experience of Zero Point. Jesus did not die for the sins of the Earth or the people upon Earth. Jesus did not die at all! In death the message would have been lost. Christ Jesus anchored the wisdom of rebirth, through Resurrection, a message very different from that of death. To those of his time not knowing or understanding the process, however, it may have appeared as death followed by the reversal of death. Jesus demonstrated that through his *Choice* of life conduct and his *Free Will* to carry out his choice, he *became* a state of awareness wherein he had outgrown the possibility of disease and death. He demonstrated, in the presence of others, that no experience may take the gift of life from an individual that has Chosen to fully honor that gift. To do so is to fully empower our possibilities of humanness.

Jesus was executed because the consciousness of Earth allowed his execution, the result of intolerance for his message of empowering love. The innocent consciousness to whom he had come to deliver the tools of completion, those that he loved so much that he descended into dense matter to guide them toward their own completion, were the very souls that provided the environment, setting the stage for the events that followed. It was those individuals, in their ignorance, through attempting to "kill" Christ, that effectively sealed his message of love, Compassion and forgiveness into our memory, so that we may access it today. Using their tools of Choice and Free Will, expressed as fear, guilt and ego, it was the peers of Jesus that catalyzed his ability to perpetuate the message that he had come to deliver. Through his execution, Jesus demonstrated that there is no "death," only a change of expression; what physicists would today term a "change-of-state." Jesus reappeared whole, complete, healed and very much alive, as a Resurrected being in complete resonance with a higher expression of creation. He reminded us of our truest nature as "angels" of light.

The gift that Christ Jesus offered to this world was the anchoring of his awareness firmly into the conscious matrix of all humans present at that time, as well as those yet to come. Through his womb-birth and life experience, others saw, felt and experienced the love of Christ Jesus in their own lives. His execution and subsequent resurrection served then, and now, as a *living bridge* between the experience grids of every day, and those of a much higher expression of the same matrix; the grids holding the possibility of Christ Consciousness.

Through the use of Choice and Free Will in the manner in which he conducted his life, Jesus demonstrated that resonance to either grid is not only possible, it is also acceptable. He lived the life of man as a stepping-stone to becoming a fully empowered human. Christ further demonstrated that there is no judgment from creation itself against any one individual for actions or deeds resulting from the use of Choice and Free Will. There is no day when the weight of "sin" will prevent an individual from moving into a higher form of expression or "heaven." There is choice, and the consequence of choice, expressed and mirrored as the patterns of our lives. Perhaps the greatest factor of limitation within any lifetime lies within us and the way that we judge ourselves through the consequences of our choices.

Repeatedly, ancient and indigenous traditions remind us that we determine our destiny. We determine how, and to what degree, we progress along our evolutionary path, moving past our illusions of limitation, to the freedom that lay beyond. We create the opportunities and choose our paths rather than living the plan formulated by a high panel of regulatory beings. Certainly highly evolved beings, guides and Ascended Masters are present throughout our life-path journey. We are reminded that the primary tool of our journey, however, is life itself. The choices that we make on a daily basis are the elements of this tool. How do we feel about ourselves? How do we feel about other individuals? How do we treat others in day-to-day relationships; on the street, in the stores or office place? From this perspective we are probably less concerned with the outcome of each choice as we recognize that within each choice lies an opportunity of mastery. It is through the very process of the rich and varied experiences that we create the eventual outcome of our evolutionary path. The manner in which we choose to conduct ourselves outwardly provides a mirror as to the degree of resonance attained within, to higher grids of Compassion and nonjudgment.

Life reminds us how to "unplug" from the old grids harboring beliefs of hate, fear and separation expressed as the laws, the rules and dogma that were ingrained into our memory over periods of thousands of years. As we allow for new possibilities of expression, we "drop" the old programs that may no longer serve us (Fig. 67). Our ability to accept positive change, our openness to new ideas and concepts, our willingness to change an old pattern that no longer serves us in exchange for the opportunity to create something new to fill the void are the manifestations of "unplugging" from old grids. To the degree that we are able to *allow* new patterns in life we find a strengthening of our connection with the new. These are signs of a new wisdom. Our thoughts, beliefs and the manner through which we express those thoughts and beliefs are our tools of change; our keys to the grids of "Christ Consciousness."

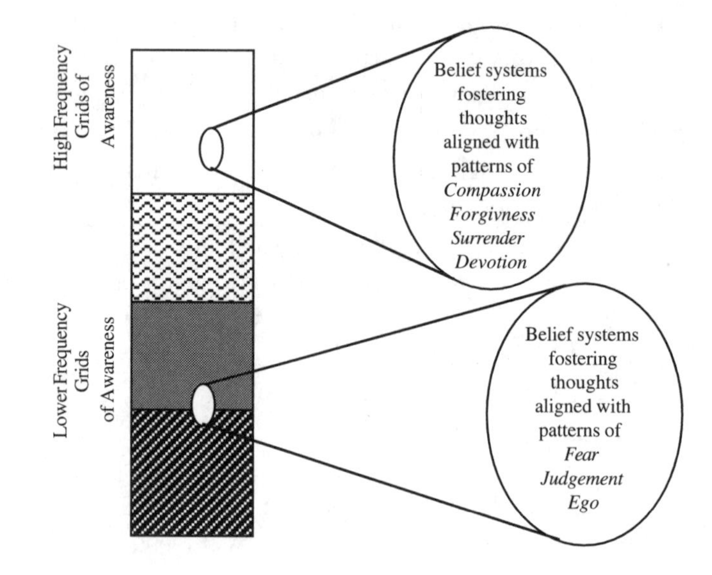

Figure 67 *The beliefs and thoughts of life are the tools that allow us to "unplug" from grids holding old belief systems and values. The use of these tools provides a path toward new bodies of information aligned with the energy of Earth's evolutionary path; the grids of Christ Consciousness.*

It is to these grids of possibility that Christ Jesus "anchored" the bridge that has made this information easily accessible within the present lifetime. It is for the purpose of building this energetic bridge that Christ descended to this world, lived, was crucified and resurrected nearly 2,000 years ago.

This was the gift of Jesus, our message of eternal life through conscious use of thought, feeling and emotion expressed as daily life conduct and our use of will, intent and prayer. This state of awareness is achieved by intentionally shifting our focus of viewpoint, resulting in molecular shifts within our bodies toward greater expressions of themselves. We call this vibratory technology of awareness Resurrection. Jesus demonstrated his gift through offering his truth in a world that was not openly accepting, anchoring divine wisdom, bridging the old to the new; the path to the Christing of Earth. We live that Resurrection today as the choices in life.

THE TIMING OF JESUS' MESSAGE

Through his living example and teachings, Christ Jesus provided a reference point for every individual, as a reminder of the possibilities of expression within a given lifetime. His gift was a message of hope, remembrance and empowering love. He was welcomed with mistrust, disbelief, fear and nonacceptance. In addition to his message being misunderstood during his lifetime, today the interpretations of Christ's teachings remain a source of debate and controversy. This is especially true for those relying upon modern interpretations of the ancient Biblical texts. Acknowledged by historians and scholars alike, our modern versions of Biblical texts are nonsequential and incomplete representations of a full and rich body of knowledge, recorded in some cases hundreds of years after the actual events themselves. Though available, and in some cases more complete, many ancient texts are becoming inaccessible to present-day seekers for a variety of reasons.

The Dead Sea Scrolls, for example, have been recovered, restored and translated. Now legal precedents are assuring that many will remain "closed" to the general public, with access granted to a select few. In the summer of 1993, the Jerusalem Court awarded the copyright of a particularly significant text, the 2,000 year old MMT scroll, to Elisha Qimron of Ben Gurion University of the Negev.[5] Without these scrolls, and other texts including the original Biblical books of Enoch, Protevangelion, Christ and Abgarus, Nicodemus, Laodiceans, The Apostles Creed, Philippians, Philadelphians, Romans, Trallians, The Letters of Herod and Pilate, Hermas and Magnesians, tangible records of the teachings of Christ appear to remain inaccessible.

Though the actual words of Jesus' message may not be seen directly, the patterns of light, the vibration anchored through Christ's life are both available, and accessible, to those choosing to allow themselves resonance to the memory of those truths. Why did Christ appear at the time, and in the manner in which he did? What was his intent? Was he successful? A schematic view of Jesus' appearance on Earth in relation to the overall length of the cycle provides insight into the timing, as well as the geographic location, of his point of entry.

Earth, today, is living the completion of a cycle that began nearly 200,000 years ago. Another way of describing the timing of Christ's birth 2,000 years ago is that it was 2,000 years *before* the end of this cycle, 2000 A.D. An event occurring 2,000 years prior to the completion of a 200,000 year cycle indicates that approximately 99%, or 198,000 years of the cycle, has completed (Fig. 68). Note how much of the cycle is to the left of the "0" mark indicating the birth of Christ. The intent of his emergence onto the Earth plane was to serve as a reminder of the evolutionary process that is at hand. With 99% of the cycle completed, it appeared that this would be an optimum time for this reminder. Too early in the cycle and the message would not be understood. Too late in the cycle and consciousness would not have time to assimilate the teachings.

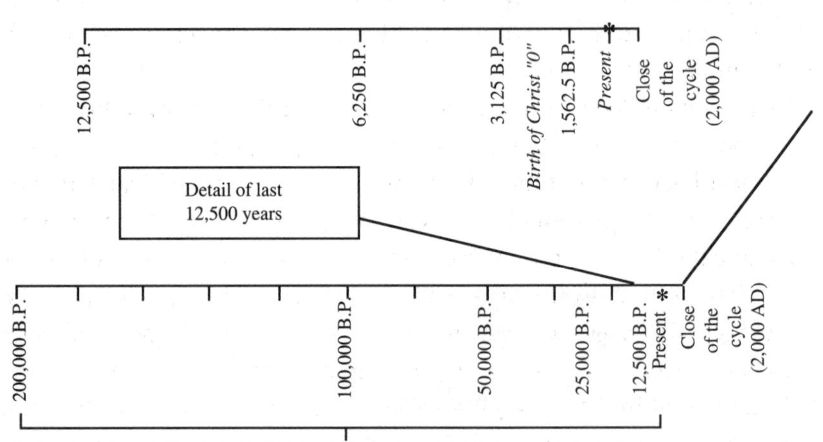

Figure 68 *Timeline showing the relationship between the beginning of the present cycle of awareness, the birth of Christ and the close of this century.*

Consideration of the relative intensity and location of global magnetic fields provides yet another indication as to the timing of the birth of Jesus in relation to planetary cycles. You may recall that, in general, relatively lower magnetics provide a greater opportunity for change, while higher values of magnetics provide a buffer of interference and less of an opportunity for change. Figure 10 in Chapter 1 indicates that Jesus chose a time of relatively high planetary magnetics to introduce his message. Why? Why would he not have chosen a time in history when Earth would be open and accepting of his offering and message of hope? Why did he choose to anchor his message in the Middle East?

To answer these questions, it becomes necessary to reexamine the intent of Christ's descent to Earth. His gift was in his life; the truth that he lived in the face of a world that was unsupportive of his message. To firmly anchor his message it was necessary for him to model his truth, to demonstrate the power of his faith and love for Earth, in a manner that could not be disputed in his time. He used the actions of those who did not support him to work in his favor, by allowing the amplification of his message through their own fear. Their actions proved that the message of Jesus the Christ was valid, as they forced the test. This was precisely the opposite of their intent to silence him through death. To have offered this message in a world of low global magnetics offering acceptance and lenience, the gift would have had neither the potency, nor the lasting effect.

Interestingly, though Earth was experiencing generally high values of magnetics, a contour map of global magnetics indicates that the Middle East was, in fact, experiencing relatively low magnetics (see Chapter 1, Fig. 11). The net effect of these parameters may be seen as follows:

For the time in history that the universal reference being of Christ Jesus chose to deliver his message, he chose the best possible place on Earth with a high population density, relatively accepting of new ideas and relatively low magnetics, in the knowledge that these same conditions would seal his fate as a man, while anchoring his message into the human matrix.

There appears to have been an assumption regarding the evolutionary nature of human awareness. The assumption is that conscious progress is made more or less uniformly in a linear fashion over time. That is, while not every individual within each race would arrive at the same plateau of understanding during the same period, humanity as a whole would experience a uniform increase in the conscious aspects of awareness.

Throughout our evolutionary cycle at various "milestones" of progress, there have been surges in our curve-of-memory as reference beings anchored concepts of oneness, in the absence of separation. Conscious evolution, defined by the critical mass of thought required to trigger resonance with the next level of awareness, is evolving in accordance to a geometric rather than a linear curve. Appearing as a slow progression over long periods of time, the process is augmented with sudden and abrupt increases in concept and thought. Composite awareness is evolving so rapidly at the present, that the ancient prophecies predicting selected individuals surviving the experience of The Shift into the new reality deserve reexamination. Each human living within the envelope of Earth's magnetic fields at the time of The Shift will experience the process of dimensional translation—in Biblical terms the Resurrection and Ascension. All will have a *choice* as to how they experience the process; consciously or unconsciously, awake or asleep. This was the message that Christ brought to this world, and with the message the tools of conscious Resurrection and Ascension.

THE MYTH OF SIN

Ingrained into the matrix-memory of the Earth is the belief that each individual blessed with the opportunity of a physical experience begins from a lesser state of spiritual development; a state attributed to having "fallen" from a more evolved level of evolutionary progress. The term describing life-choices resulting from a fallen state of grace is that of "sin." While modern texts define sin as "a transgression of a religious or moral law,"[6] ancient texts refer to sin simply as the sense of "separation."

In the density of magnetics and extremely low frequencies (ELF) of our Earth experience, we may well have experienced a sense of separation from all that we had known prior to our "arrival." From this perspective, "sin" is viewed very differently from our modern definition of a transgression. To extrapolate our "separateness" to the premise that our spiritual fate is predetermined to be that of a lesser being by virtue of our very birth makes little sense. Historically, our conditioning asks us, individually and collectively, to redeem ourselves from the wrongdoings of life in the eyes of our Creator. Additionally, we are reminded that regardless of our chosen path, it will not be possible to attain the degree of spiritual evolution demonstrated by our universal being of reference. Through the traditions of our own ancient texts, those of the Essenes and the oral traditions of our indigenous ancestors, we are offered a very different life view. Following is a partial summary of common misconceptions perpetuated through distortions of the original teachings:

- We are born into sin and have the opportunity of our Earth experience to redeem ourselves in the eyes of our Creator, approaching while never reaching the evolutionary state of the Universal Christ.
- We are fallen angels, flawed from birth by virtue of the fact that we were born into the Earthly experience.
- Because we are fallen, we will require an intermediary, someone to act in our behalf, to intercede for us, with our own Creator.
- Our life-plan is a mystery to us. The course of our fate is predetermined by a plan that we are incapable of comprehending.

Today, as the close of this cycle nears, unprecedented numbers of individuals are turning away from the "traditional" belief systems based in these distortions. Though the specific reasons may vary; in general each is discovering through his/her unique experience that the relatively modern religions, based in fear, ritual and dogma, do not provide the tools to address the challenges they are asked to face on a daily basis. Traditional beliefs do not serve them as they are faced with the unprecedented challenges inherent in the shifting energy of the close of a cycle; their own fears, "failing" relationships and waves of emotion that have never been experienced before. Belief systems based in distortion cannot meet their needs, because they are rooted in the fundamental premise that we are helpless and powerless—unable to influence the outcome of events within and outside of ourselves.

It is for these reasons that today, in the closing years of this cycle, so many search to find a more meaningful expression of what is felt within. The search has led them into the "nontraditional" belief systems of the ancient, indigenous and forgotten peoples of the Earth. Though the words may be different, there are underlying threads of continuity within the ancient traditions and the actual teachings of our Universal Christ, rather than the interpretations of the translated and fragmented texts seen today. Among those threads of truth, consistent within the Egyptian, Native American, Buddhist, Tibetan, Essene, early Christian and ancient mystery schools are the following fundamentals:

- We are a part of all that we see, with the opportunity to live in harmony with creation, rather than controlling and ruling creation.

- We are more than "fallen angels." We have arrived in this world by choice, having consciously chosen to descend into our Earth experience for a specified period of time.

- We are extremely evolved, powerful and masterful beings; the creators of, and experiencing the consequences of our patterns of thought feeling and belief.

- We have direct access to creation through ourselves. We are "sparks" of the divine creative intelligence that is responsible for our very existence.

We are, and have always been, equal to our angelic counterparts, viewed as such by all except ourselves. We are truly the creator of our world(s), and as such we are a part of all that we see and all that has ever been. We are the Alpha and the Omega—the beginning and the end, all possibilities existing as potential; awaiting the opportunity to coalesce as our thoughts, emotions and choices.

> *You are as new as the moment you were shined into life from the heart of God. Your radiant self has not been touched or tainted by Earthly experience or any darkness you have perceived.*
>
> THE PEACE THAT YOU SEEK[7]
> "HOW PERFECT THE WAY," ALAN COHEN

COMPASSION AS EMPOWERED EMOTION

The time of our outer temples, outer networks, outer grids and external guidance is nearing completion. For many, what has come as an inner knowing was stated clearly through the language of the time two thousand years ago, and even before. Our knowing reminds us that we live as an expression of a highly sophisticated union, *a sacred marriage*, between the elements of this Earth and a directive nonphysical force. We call that force "Spirit." Perhaps the most astounding of the ancient references restating its validity today is the link between feeling, thought and their relationship to human physiology.

> *Three are the dwellings of the Son of Man, and no one may come before the face of the (One) who knows not the angel of peace in each of the three. These are body, thoughts and feelings.* *(Parentheses are the author's.)*
>
> ADAPTED FROM *THE ESSENE GOSPEL OF PEACE*[8]

The peace that we seek in our world and in our body is the same peace of this Essene reference. Compassion is defined as a quality of thought, feeling and emotion. Compassion may be demonstrated as a quality of conduct in our daily lives. The vitality of our body, the quality of our blood and breath, our choice of relationships and emotion, even our ability to reproduce, appears to be directly linked to our ability to embrace the force of Compassion in our life.

To the degree that we embrace Compassion in our lives, change passes gracefully, with ease. For those who require proof, that proof is now available. For others, simply knowing that there is a direct relationship between emotion and DNA comes as welcome validation for an inner knowing that has driven the course of their lives for years.

Following our definition of a science, if we do these things, then such-and-such will happen, clearly the ancients left to us a path. Today their path may be viewed as a science chosen to carry us gracefully through The Shift of the Ages. Our word for a sometimes nebulous state-of-being, Compassion is a quality of feeling, thought and emotion allowing the 1.17 volt liquid crystal circuitry within each of our cells to align with the seven-layered liquid crystal oscillator within our chest that we call "heart." Compassion, the result of coherent thought, feeling and emotion, is the program that you encode, determining your body's life giving response to the reference of Earth's heartbeat. Beyond simply feeling, Compassion is the merging of feeling with emotion and directed thought made manifest as our bodies!

Compassion is the kernel of your very nature. The science is offered as a program of language and understanding as follows:

AS
we allow life to show us ourselves in new ways
so that we may know ourselves in those ways

AND
we reconcile within ourselves that which life has
shown us

THEN
we become Compassion.

It is within the very reconciliation, the coming to terms with whatever we have invited as life, that we become Compassion. Deceptively simple, the understanding of life's mysteries has been the subject of controversy and debate for centuries. To what extremes have we taken ourselves to know of the darkest of the dark and the lightest of the light? Clearly, the ancients remind us of two things:

- That the events of life serve us by allowing the opportunity of feeling and emotion through a broad range of experience: all of the "good" and all of the "bad."
- Further, there is a pattern to the order within which we will recognize the experiences: There is a sequence and a progression to the experiences.

The keys to Compassion, then, lie in our ability to embrace all experience as part of the One without judgment. To live solely in the "light," shunning, ignoring, working against and judging anything other than light is to defeat the very purpose of your life in a world of polarity! It is easy to live in the light, if the light is all there is. You, however, have come to a world where light exists in union with its opposite.

Have we fallen into the ancient trap of deception where we:

- See one aspect of polarity as better than another?
- Believe that one aspect of our world of polarity is of something other than the Creator?

I often hear of individuals who view themselves as spiritual warriors fighting the battle of light and dark, drawing the spiritual lines of war. This perspective is a path. Each path carries its consequence. Inherent in the perspective of battle lines lies judgment, the very hallmark of polarity.

There can be no battle without judgment.

In a world where we have come to experience and know ourselves in all ways, how can there be a "right" and "wrong" in the experience itself? It is the assignment of good, bad, light and dark as judgment that implodes unity into polarity. Is it possible that darkness is a powerful catalyst in our lives, similar to the viruses discussed earlier, catapulting us beyond polarity into an even greater technology born of Compassion?

Possibly the greatest honor that we may bestow upon ourselves, as well as the greatest challenge that we may encounter in moving toward great states of personal mastery, is to return to our truest nature of Compassion. Compassion may be discovered in redefining the experiences of life for what they are, rather than what our judgment and bias will make of them. To become compassionate within life's experiences is not an invitation to callousness, void of feeling and emotion. Quite the opposite, it is through the very act of allowing ourselves feeling that we are guided toward those portions of ourselves seeking the greatest healing. Feelings and emotions are our tools from which to access the reasons underlying the intensity of our emotion.

Within each cell of our body exists a "bioelectric" potential, created through differing charges of fluids on either side of the cell membrane. Our brain regulates these potentials by maintaining an acid/alkaline (pH)* balance. When we feel or emote in response to our life's offering, we are literally experiencing the shift of electrical charge from one place in our bodies to another! Our feeling in that moment is actually the shift of the pH balance within our brain and the resultant shift of electrical potential across cell membranes throughout our body. Emotion is what the electrical charge within the liquid crystal of our body feels like.

It is this shift that polarizes our experience into a positive-negative-good-bad relationship, effectively preventing a neutral or balanced viewpoint. As our electrical potential shifts, we broadcast into our world the frequencies of that which we continue to have a charge upon. These are our judgments. Drawn by these frequencies, we will attract into our reality those individuals, circumstances and events that will provide the greatest opportunity to witness our judgments mirrored to us through the actions of others. These mirrors are detailed through the text, *Walking Between the Worlds: The Science of Compassion.*

Compassionately redefining our interpretation of what life has shown us removes the "attractors" of others that mirror our greatest potentials of healing. As beings of Compassion, we demonstrate the sophisticated vibratory technology of the subtle power of thought, feeling and emotion (Fig. 69).

* A pH of 7 is neutral. An increase indicates greater alkalinity. Decreasing from 7 indicates greater acidity.

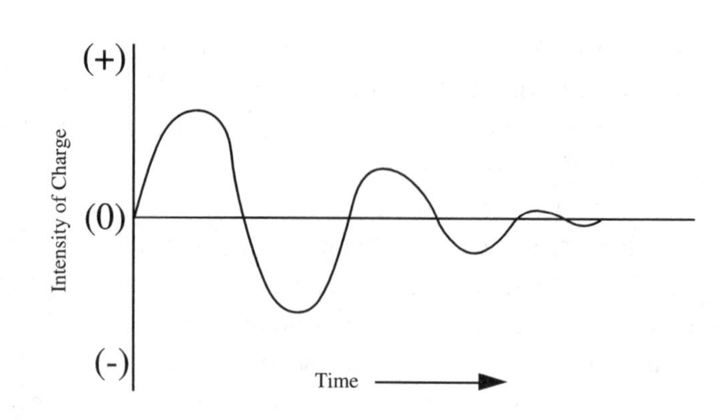

Figure 69 *Schematic diagram of charge associated with judgment. Compassionately addressing life's challenges compresses the time required to dissipate the charge.*

Interactions with others, especially now in this time of psychic fine-tuning, when there is a tendency to empath, may appear to be "emotional roller coasters" in the absence of Compassion. We see why Compassion cannot be falsified. Masking the *outward* emotion alone cannot alter the *inward* pH shift of the body. Our bodies recognize our truth on the level of the cell. For this reason, close and honest scrutiny of relationships, acquaintances and situations will yield direct insight into your present state of bias and judgment.

Within ancient temple sites, such as the King's Chamber in Giza, the towers of Southern Peru and the kivas of the American desert Southwest, are found structures that provide an environment within which we may access ourselves within a zone of neutral charge. Through the "passive dynamics" of the structure's geometry, magnetic fields (the glue) binding the grids of fear, judgment, bias and ego are significantly decreased. The decrease in magnetics mimics similar processes that may be generated through specific thought streams, such as those taught through the mystery schools and Zero Point Meditation. The net result of the process is direct access to the electrical essence of self in magnetic conditions very close to null: *Zero Point conditions.* Though effective, these are examples of an external technology intended as a tool, to induce an inner experience of Zero Point. Though valid, the tools are probably not necessary. Jesus demonstrated our gift of compassionate technology, the tool, with which to accomplish the identical process from within.

One of the greatest, and possibly least understood, messages that Christ Jesus anchored by living it in our presence was that of demonstrating love through compassionate allowing: loving others enough to allow the latitude of their experience. This becomes possible through the *knowing* that each comes to Earth in a unique capacity, from varied backgrounds and modes of expression, and that all are equal in their response to the challenges of life. To the degree that any aspect of another's experience is judged, to that degree do we remain in the polarity of separation and the charge of that judgment.

Compassion may be demonstrated as a quality of conduct in our daily lives. Though the vocabulary, culture and society have changed, Jesus' message to us is just as valid today as it was 2,000 years ago. It is a message of inner technology. Our inner science is the root of, and supersedes, all that we may engineer and build outside of ourselves. Our world is a mirror of processes from within, tools that we create *to remind ourselves of ourselves*. Through his life and Resurrection, Christ's experience was a metaphor for our lives. Through his execution, he modeled for all of humanity a process that each will go through individually, as well as collectively. In the closing years of this cycle, every life form existing upon the Earth will have the opportunity to experience the dimensional translation as Earth's shift from a third to a fourth density experience; a Zero Point awareness and the choice of Resurrection over death.

THE ETERNAL FORCE

Throughout the many and varied facets of human history, there has remained one constant expressed as a subtle yet powerful force, driving the process of life forward. This force is experienced within each individual as the sheer will to continue life. Driving each individual, as well as the collective whole, life moves forward toward some goal, some point of resolution. To many, the resolution is anticipated as a "something," some event to make the experience of life worthwhile and complete. It is this something that has provided the momentum for consciousness; the inertia of will that perpetuates the experience. Somewhere within the remote memories of human past, the meaning of this force kept itself alive as flashes of insight. Sometimes appearing as glimpses of divine insight, we have been reminded of our relationship with this force through dreams and familiar emotion. A message of some kind, pierces through what feels like a barrier into the awareness of the present to remind humankind of itself, its meaning and purpose. Evidence of that force may be seen throughout history as the relentless search for truth and knowledge, a search consisting of many lifetimes at the cost of many lives. What could be the foundation of such a desire? What drives our search to know and remember?

In and of itself the knowledge of life offers little meaning. Without context our life experience may appear as incongruent streams of information to be stored and accessed at some later date. It is the living of our knowledge that becomes our expression of the search. This is the wisdom that provides meaning to life. Within the context of our wisdom we are compelled to remember life's purpose in our own terms. From that purpose, from our knowing, stems our reason to continue. We feel on some level, that we are approaching a time when something within us will change. We sense a time when all of our knowledge gained throughout all of our lifetimes will apply. Some describe this feeling as a sense that they have been in training all of their lives, preparing for "something" really big.

The outward expression of these feelings historically has appeared as inconsistent, sometimes unfathomable, extremes of experience. The reality is that each of the six billion (plus) humans incarnated at present, through unique and individualized experience, is expressing precisely the same pattern of energy. This is the mysterious irony. Though there are many bodies in our world, there is only one of us here! We seek the memory of ourselves. As a single consciousness we must experience our extremes to find our balance. Each life experience becomes another lens from which to view the wonder, mystery and possibilities of life. Each individual fills a particular niche, purely through the uniqueness of themselves, within which to express the feeling of a shared and *awesome force*. What is this force?

The energy, driving us toward a point of resolution, is what many call our "will" to continue, our will to return to the "One." The force is the force of life, life force, Chi, Ki or Prana. By virtue of its very nature, life is compelled to reassemble the fragments of itself, time after time following each experience, and return to the wholeness of pre-experience. Will is the name given to the patterns of consciousness striving to attain this state of balance and completeness. Within our third dimensional world, will is expressed through our bodies, our liquid crystal resonators within which spirit may resolve itself. Our force always has a message and our message is always the same. In its consistency, the message becomes a truth, an absolute and universal message rather than a relative truth that works under some circumstances. Will reminds us that there is a force within each of us, that is infinite and eternal, and may not be created or destroyed. Life is our expression through which we remember that force.

Before continuing, I invite you to stop and consider the statement that you have just read. You have heard it many times. Do you know it to be true? Do you feel it to be true in your world? Do you believe and have faith that it is so? We have often confused the experience of life with the energy of life essence. The electrical patterns of our life essence remain and continue, regardless of the outcome of our life experience. Through the gift of life, however, we each have been given an opportunity to express our essence uniquely, perhaps in ways that have never been accomplished before! Pause for one moment and allow the impact of the simple statement above to become a part of you. Should you find yourself in disbelief of this statement, you may want to ask why? What it is that has occurred in your life to "teach" you not to believe a universal truth?

From the perspective of the ancients and the teachings of the indigenous peoples today, there is no real life, nor real death. The totality of our experience is that of a dream. Within that experience, life and death are "dreams within dreams"; with no separation between the two. There is only the perception of different and varied experience and how we think and feel about those perceptions. Consider precisely the same truth presented within the context of another language; a language that accesses another portion of our wisdom. Within the conceptual framework of Western science, life force may be thought of as energy-light-information, an array of electromagnetic pulses arranged into orderly patterns of experience. The language of our science reminds us that energy may not be created or destroyed. Energy responds to varied experiences by changing the manner in which it expresses it's form.

The energy of life force permeates all of creation, transcending the boundaries of dimensionality, space and time. Life force is essentially infinite in nature. The now-famous equation of Einsteinian physics, from earlier in this century, equates mass to energy and states this truth very explicitly: $E = MC^2$ where E = energy, M = mass and C^2 = the velocity of Light multiplied times itself. This equation, in an equally valid language of mathematics, states that as matter accelerates, its mass begins to expand.

Upon approaching the speed of light,* the mass expands to the homogeneous state described as that of "infinite." Matter as a "group" vibration, accelerating into another "group" vibration, to become a new expression. We call that new expression light.

Two very different languages, expressing precisely the same truth. One language is very intuitive and right-brain oriented. The second more analytical and left-brain oriented. Both are equally valid. It is this truth that drives our will, our desire to attain a state of development that allows eternal life. We remember this truth within ourselves. We know that our essence may not be created or destroyed. We have to have known this to have navigated ourselves through the matrix of creation, propelling our essence into the crystalline form that we call "body" for the experience of life!

> *No man can reveal to you aught but that which already lies half asleep in the dawning of your knowledge... And even as each one of you stands alone in God's knowledge, so must each one of you be alone in his knowledge of God and in his understanding of the Earth.* THE PROPHET, KAHLIL GIBRAN[9]

Each individual must come to terms with this truth within the context of their own life and the uniqueness of their experience. To give the gift of yourself, in wholeness and completeness, you must know yourself within the context of all possibilities, all the extremes. All of your joy, all of your pain, all of your anger, rage, jealousy, and judgment; those precious feelings are your gifts to help you to know yourself. Your unique experience allows you to push the boundaries of who you believe you are, as you approach the reality of what you are truly made of. In that knowing, you have the opportunity to see yourself in situations that you may never experience again. These are the extremes that will help you to know and redefine your point of balance; they are constantly shifting.

Though it may be masked within the distractions and rationalizations of our day-to-day world, it is the indestructibility of our soul that provides the momentum to get up every morning and go on, moving forward with life. We "know" of the eternal nature of consciousness on a deep level; it is encoded within the light patterns of memory that reside within each cell of our bodies. Eternal life is based within eternal truth, an absolute mirrored within each cell of our creation matrix. Our bodies appear as whole and complete unto themselves, expressing within each cell of existence. At the same time, our individual patterns are part of a much greater whole; the pattern continues. This, then, is the holographic law. Our life force is eternal, holographic and recursive in its existence. Preservation of consciousness is the law of creation. How may that which is truth become unmade?

This truth is the message of our ancient prophecies. This foundation provides the basis for the texts dating before history. This is the common thread linking each religion, sacred order, sect and mystery school. The essence of who we are, in the absence of fear, judgment, ego and any other distortions of our truest nature, is eternal.

* Recent studies indicate that the speed of light is not a constant, with electrical signals recorded traveling at over 100 times the accepted speed of light (186,271 miles per second).[10]

The world that we have created for ourselves, our lives, family, friends, and career, our surroundings even our behavior patterns result from our feelings and beliefs; temporal patterns of energy that we have created to serve us within the context of our life blueprint, to see ourselves from many viewpoints. All viewpoints are in support of this time in history, when the sum of experience focuses upon now; the transmutation of ourselves and Earth into a more refined pattern of expression. The prophecies remind us that this process is our purpose; to remember the possibilities of a new expression of the creative frequencies of our life.

Jesus reminded us of our possibility through his words, "I am the resurrection and the life; He who believes in me, though he die, yet shall he live, and whoever lives and believes in me shall never die."[11] The Shift of the Ages is resurrection and new life, born of a new wisdom! We are asked simply to allow and remember.

SUMMARY

- The gift of Christ Jesus to Earth was in his life rather than his death. Jesus never died, he Resurrected! Resurrection is birth from an existing life rather than life's end followed by new life. It is the message of the continuity of life that is celebrated. 20th century science has demonstrated that energy cannot be destroyed. It may, however, change the form of its expression, and that is precisely the process that Jesus modeled to Earth nearly 2,000 years ago.

- Earth is evolving to a new form and we have the opportunity to accompany Earth into a greater expression of itself. Our essence is that of life force; energy that transcends creation and destruction. Only the form, the manner in which the energy is expressed, is changed. The change is a direct function of vibration.

- Life is the tool that allows great shifts of vibration. We seek a condition of experience allowing the latitude of life in the absence of judgment. The words that express this state of being are those of "Compassion," "allowing" and "love."

- To the degree that we label an aspect of our experience as "sin" and see the experience as separate from ourselves, to that degree do we remain in the illusion of separation, polarity and the charge of judgment.

- Ancient words worth remembering, and well worth repeating, are those of the Universal Reference being who lived the words: "I am the resurrection and the life; He who believes in me, though he die, yet shall he live, and whoever lives and believes in me shall never die."

CHAPTER SIX

From Separation To Union

THE HOLOGRAPHIC MESSAGE OF TRUTH

You have now found the conditions in which the desire of your heart can become the reality of your being. Stay here until you acquire a force in you that nothing can destroy.

MEETINGS WITH REMARKABLE MEN[1]
GURDJIEFF'S SEARCH FOR HIDDEN KNOWLEDGE

It was the Spring of 1994 that signaled the beginning of what the locals call the "dry season" in the high Andes Mountains of Peru. During the third day of a four day trek, my group of 22 hikers, five Peruvian guides doubling as cooks and 22 porters, would cover approximately seven and one half Peruvian mountain miles. Crossing over three mountain peaks above 12,000 feet, we would descend from each pass into the lush, green cloud forests 3,900 feet below in preparation for the ascent on the next pass. The evenings had been well below freezing and my goal was to get each of the trekkers into camp and warm, dry clothes before the temperatures dropped dangerously low. We had just become aware that during the same time one year previously, two of the porters supporting a similar trek had died after the second day, having frozen to death sometime during the night. Wet clothes and subfreezing temperatures are ideal components for hypothermia, a condition in which the body loses heat faster than it is able to generate heat.

Though we had acclimated for five days, illness and altitude had weakened some of the hikers. Our group had essentially split into two groups, separated by several miles, each being led by expertly prepared Peruvian guides and porters into a camp with warm tents, hot meals and tea awaiting them. Hiking through the morning hours with the lead group, we arrived at a temporary lunch camp with freshly prepared meals of fresh bread, avocados and tomatoes. Following a light meal, I chose to double back, checking on the distance and conditions of the second group. They had been joined by a Peruvian holy man who, along with our guides, would accompany them into camp. Satisfied that they were together and in good hands, I headed back toward the first group, already well on their way to the evening camp.

As I climbed the steep, rocky trail I looked toward the summit of the pass above, and glanced down the talus slope behind me. Suddenly, I realized that for the first time since we had left Miami six days earlier, I was alone; absolutely and completely alone. As I neared the pass, I paused briefly to immerse myself in the sheer beauty of this land. Although there were still several hours before dusk, the sun was already dipping behind the peaks towering over the valley below. We would all be hiking in shadows soon.

Directly below me was a glacial lake that I had not noticed before, like a perfect crystalline mirror, reflecting the high peaks that surrounded me. The rich and intense colors surrounded me in all directions. Deep emerald green jungles below supported the snowcapped peaks jutting into that intense, blue and crystal clear sky that we always thought photographers somehow doctored up for the magazines. A gentle breeze brushed my face, welcome relief from the howling winds that had ripped through the valleys just hours before.

Silently, I gave thanks for the opportunity to experience such raw beauty as I completed the short distance to the saddle of the pass. With the top just a few feet away, I paused sitting on a smooth rock that looked like it was made for just that very moment. I had carried a wooden Native American flute from New Mexico to accompany our group prayers and meditations. This pass felt like a perfect opportunity to offer a melody of thanks. I pulled the flute from its protective cover and offered long, slow notes. Quickly, each flowed into the deep, resonant tone of a melody coming from somewhere within me. It was a melody that I had never heard before. The notes echoed from the rocks in front of me as the wind carried them away. I remember wondering if the others could hear my song. They later said that they did not. I began to breathe deeply, inhaling the sensations of one of the purest places on Earth.

Suddenly, quite unexpectedly, I felt tremendous waves of emotion surging through my body, pulsing from my chest outward. The pulses grew stronger and the waves more intense. Tears welled up in my eyes and I found myself crying uncontrollably in long, deep sobs of appreciation for this moment of sheer beauty. As I wept in the experience, I noticed a shift in my body. Once again, it was the *feeling*, surrounding me, engulfing me in the warmth of surrender and allowing. This was the feeling, the familiar feeling, from Lost Lake 17 years earlier. Though I had often created the feeling willingly, this feeling was simply happening, quite spontaneously.

One of the reasons for offering this four day trek as part of the Sacred Journey to Peru, is for the pure physicality of the experience. The effort exerted to complete this journey takes so much energy, that there is nothing left to stand between each participant and his or her emotions. There is nothing left to hold the "walls" of distance, distraction and indifference in place. The trekkers are supposed to get in touch with themselves through a direct and intimate experience with creation. Now, the very experience that I had intended for the other hikers was happening to me, there on a 13,000 foot pass, unplanned and unorchestrated.

As the familiar feeling surged through my body I closed my eyes and experienced the connection, the absolute and complete resonance with the creative forces that have always been there for me. In the days prior to leaving for Peru, and throughout the journey, there had been a question that I had chosen to resolve. The question embodied a core understanding vital to the very process of my life. In the space of this acceptance and loving resonance, once again I asked the question.

Father, I ask for the wisdom to know of the relationship between the forces of light and dark. Please guide me to understand the role of these forces within my life so that I may know of their resolution.

The wind picked up and began to dry the tears that had fallen into my beard. As I wiped the salt from my eyes, I perceived a voice, a familiar voice that I had heard many times before. Again, the voice was neither male nor female, was from nowhere specifically and everywhere at once, and began with a single question, to me:

"Do you believe in me?"

Without thinking, my body responded with a mental *yes*. The voiced asked again.

"Do you believe that I am the source of all that you know and all that is your experience?"

There was no need to think or ponder the answer. I had affirmed many times before in prayer my belief in the single source of creation, the fundamental vibration, the seed tone of the standing wave that allows the hologram of life's patterns. Without thinking, once again my body responded with a mental *yes*. The voice echoed its reply.

"If you believe in me, and you believe that I am the source of all that is, then how can you believe, at the same time, that anything of your experience is other than me?"

Upon hearing the words, a tremendous sense of resolution filled my body. Though a part of me had always known what the words had just offered, for the first time, I felt the wisdom of that truth. My body actually felt the knowing. Of course there were still the words of light and dark in my vocabulary. Those words would never mean the same thing to me again. Specifically, the conditioned belief that darkness is a force, unto itself, a fundamental power, in opposition to and separate from all that is good, no longer held any truth.

Each of us is asked to reconcile the force of darkness each day of our lives through the direct experience of one of darkness' many derivatives. Fear, anger, rage, hate, jealousy, depression, control issues and violation are each expressions of darkness playing out in our modern lives. In that moment, through the tool of my own logic, I was offered the opportunity to recognize darkness for what it is rather than what my conditioning had taught me. I was able to see darkness as a portion of the whole, a part of the source of all that is, rather than a fundamental force to contend with.

"If you believe in me, and you believe that I am the source of all that is, then how can you believe, at the same time, that anything of your experience (including darkness) is other than me?"

Through our perceived polarity of darkness and light, we have the opportunity to view ourselves from a different perspective, a necessary perspective, if we are to know and master ourselves in all ways.

The entire event had occurred in less that 15 minutes. In less time than it would normally take to eat a meal, at an elevation of 13,000 feet in the Peruvian Andes, I had felt an experience that would forever change the way in which I approached the collective perceptions of light and darkness. This mountaintop experience demonstrated to me, through my own logic, that darkness and light are not two distinct and separate forces at odds with one another. Rather, each represents a portion of *precisely the same whole,* the same source of all that is. This realization set off an entire cascade of possibilities.

We have been conditioned to view the world, and life, through the eyes that mirror separation: good and evil. *How may we find wholeness in separateness?* What happens if we change our conditioning? What happens if we remember to view the events of life for what they are, catalysts moving us into new experiences of ourselves, without "good" or "bad"?

In that moment I realized that to the degree that we "give in" to the urge to polarize our experience, to judge it as "good" or "evil"; to that degree do we perpetuate the very illusion that has separated us from all that we know, including one another! The holographic mirror assures us that we will experience that which we perceive. In that moment I felt what I only suspected before. Darkness as well as light, must be embraced without judgment as a part of creation, rather than a rival and renegade power outside of the One of all that is.

As I stood up on the rock, there was still no one behind me as yet. Moving up over the pass I began the long trek down into the valley below. I could see no one in front of me. I was still alone. Though I was in the shadows, the temperatures were unusually warm. They had not yet begun to drop. I immersed myself in the solitude and followed the loose rocks on the trail that would lead me into camp. We would all be hiking in the dark that night.

Knowledge: The First Step of Healing

For many, simply knowing, hearing the words that echo deep feelings of the heart and having a vocabulary to express those feelings, is the first step in the healing. The words mirror basic tenets, whole and complete unto themselves, that have been conditioned out of our experience. Please consider the following:

- We are the children of our Creator, however our Creator is perceived. Each of us lives as a holographic reflection of our father/mother Creator, projected through the matrix of creation into the thickened physical experience that we perceive as reality. Through the denseness of gravity and "heavy light," within the polarity of magnetics, we perceive the differences the distinctions and our illusion of separateness. We have always been one, separated in our perceptions only.

- We are of a single mind. Our consciousness forms the matrix of intelligence underlying the physical aspects of our world. It is our creative intelligence that allows us to perceive, exist, express and create in our third dimensional "soup" of creation. Extremely evolved, skilled and masterful beings, any perceived limitations stem from our beliefs of what is possible, what exists, and what we are capable of.

- We have never been separated from our source, our Creator, or one another. Stated to the positive, we have remained as one in mind, while experiencing as many possibilities of individualized expression. The only veils between us and our ancient memory are veils of perception. We may feel isolated or alone because our experience has lasted so long, and for some, the journey has been so painful. Living our lives, while allowing for the possibility of all experience, without the charge of judgment, is our living bridge to healing any separateness.

- A life on Earth is an experience of choice. We arrived in our world to experience and accomplish a goal. By agreement, our process is finite in nature and our goal is nearly complete.

Life is our path leading to our journey home, our return to Zero Point. The next days and months of our lives may well be the most challenging portion of our entire lifetimes. As our bodies compensate for discrepancies between the frequencies of where life choices have led us, and where we gracefully accommodate change, please consider that all experience is in support of our journey home. Each situation and every circumstance drawn into life is the outcome of energy we have expended to bring this healing to light.

From this perspective, each expression of anger, pain and frustration betrays a place within seeking healing. Each is based in a fear that may no longer serve us. We are more powerful than any fear that we may imagine. In fear, we express a fragment of us in our totality. In our completeness, we are of greater substance that any one fragment standing alone. We are expressions of the divine frequencies of love; thoughts created within the heart and projected through the mind of creation. These frequencies know of no dimensional boundaries and are beyond the bounds of polarity, duality, ego or fear. Perhaps our most valued tool will be the knowledge that we are not the pain, anger, or fear that we experience as we remember our truest nature. These experiences become our closest allies as we travel the path to wholeness.

The great "key" to accelerated growth may be found in allowing the frequencies of love, Compassion and forgiveness to serve us by releasing the binding charge of fear. In the release lies our opportunity to assimilate and integrate experiences based in fear into the proper grid/matrix relationships of healing. We each are whole and complete unto ourselves, while serving as a single cell within a much larger whole. We each live the holographic model, unconsciously or consciously, and our experience is a contribution to our collective memory as it accelerates toward a "critical mass" of a new wisdom. Each time one of us "remembers," we all remember to some degree.
Within the context of the holographic model:

- What happens if "seeds" are planted throughout the matrix that choose to redefine "evil," as an opportunity to view ourselves from a different perspective?

- What happens if "charge" is redefined as we remember to experience fear without judgment?

- What happens if we remember the role of service that fear may play in our lives without empowering fear through the charge of judgment?

What happens is simply this; to the degree that we redefine what "evil" and "fear" say to us as individuals contributing to the whole, to that degree do we live the wisdom offered through the ancient teachings of the ascended masters and preserved through the mystery schools. To that degree may we give the greatest gift to ourselves and to others, the gift of ourselves in wholeness and completeness.

By virtue of the holographic model, a relatively few individuals may plant the seeds, leading to a threshold of collective resonance. It takes few to crystallize the change. The Essene texts remind us so eloquently yet simply, *"The harvest is great though the laborers are few."*

SUMMARY

- Our viewpoint of "good" and "evil" as inherently separate and fundamentally opposed forces has perpetuated our illusion of separation

- Viewing "good" and "evil" as aspects of a polarity experience stemming from a common source serves as a living bridge toward healing of the separation

- If you truly believe in a single source of all that is, then how can you believe, *at the same time,* that anything of your experience is other than that source?

CHAPTER SEVEN

Awakening to Zero Point

The Bottom Line

Ultimately, you will grow beyond the teachings of the outer masters. The more of yourself that you know, the fewer teachings there will be. The fallacy of the "Path" is that there really is no path at all; there is only your experience.

THE AUTHOR

For clarity in our Zero Point perspective, no distinction is made between a "spiritual" activity and that of day-to-day life. Truly, we are beings of spirit expressing as our bodies. As such, we are capable of only spiritual activity. Each act of each day, regardless of how it is judged, what value is placed upon it, the degree of pain or pleasure that it brings to ourselves or to others, is an expression of spirit.

Many feel more at ease, however, in the distinction between activities dedicated to the spirit and those of the remainder of life. This distinction perpetuates our illusion of separateness. In an effort to summarize without separation, I have made no distinction between spiritual and nonspiritual activities. In summary of our experience:

- Earth, and our Earth experience, are nearing the completion of a cycle initiated nearly 200,000 years ago.

- The density of planetary magnetics is decreasing through the slowing of the rate of Earth's rotation. History indicates that the decrease will continue, leading into a "null magnetic zone" allowing the collective experience of Zero Point.

- At the same time, the rate of fundamental vibration (Earth's heartbeat) is increasing, moving toward a new base upon which to rest the subsequent hierarchy of vibration that enables the new vibratory structure for a dimensional "Shift."

- Each holographically tuned cell within our bodies is attempting to match the vibration of our new experience. Quicker vibration in a field of lower magnetics allows access to our emotion of unresolved experience, stored within our bodies.

- The "tools" allowing us to master our experience of healing have been modeled throughout history in the mystery schools. Modern belief systems are based in varying degrees within these earlier messages and teachings.

- The perception of "good" and "evil" as expressions of a single source of all that is, serve as a living bridge healing separation.

- The Universal Reference Being of Christ Jesus, Christ Buddha and other "Christs," in the language of their time, offered the tools of healing. The tools are just as valid today as they were two to three thousand years ago.

- Frequency generated through nonjudgment, allowing, Compassion and acceptance is that of love. Transcending the perceptions of dimensionality, time and space, as the distortions of fear are resolved, love is all that remains. Compassionate love is our truest nature and all that remains as our illusions are healed.

Shifting Through Being

From the previous discussions of our grid-matrix models of creation, the mechanism should be apparent by now as to how patterns of energy are attracted into the local distortions of the life experience, as a consequence of patterns radiated from our body-mind-spirit complex. Our patterns experienced are mirrors of patterns projected. In the "knowing" comes the responsibility of "being"; accepting that all experience has been attracted and set up by each individual as a consequence of his/her projections.

In consideration of the holographic model of our life experience, I offer the following guidelines. Each summarizes experiences of my life that are embodied within this text. The key is to simply live life consciously, fully and with intent. We are asked to choose our commitments wisely and, once chosen, to honor our commitments. Life is the expression of our intent.

Guidelines To Living Within The Hologram

- "The Patterns that we choose to have in our lives, we must first become."
  The laws of the grid-matrix relationships allow for, as well as predict, that the patterning of energy experienced is a direct and immediate consequence of those radiated and projected.

- "We will experience the patterns that we most identify with."
  Our reality is that with which we have resonance and

affinity. Other realities may coexist and be observed, without becoming our direct experience.

- "Allow ourselves to transcend that which we have been taught." To live within the boundaries of what has been taught is to limit our experience. Prior knowledge becomes the pivotal point, the springboard from which to access new and more complete truths.

- "To the degree that we are able to see the events of life through the single eye of the heart, as opposed to the polarity of the logical mind, to that degree we heal our illusion of separation." "Seeing" the events of life for what they are, rather than "looking" at the same events through the lens of what is expected, serves to heal the feelings of separation in polarity. In our heart, there is no polarity.

- "The greatest gift that we may offer to one another is the gift of ourselves, in wholeness, completeness and truth." Through living the highest truth that we are capable of, we have mirrored into the collective whole the greatest gift of all. Within that truth all possibilities exist and we become a focal point of wholeness, purely through being.

What if The Shift Never Happens?

This is not likely, as the events of The Shift are already well under way. It is unlikely that we will see the "Shift" reported on national news or in *Time* magazine. There will be those who never acknowledge that it has occurred. If you find yourself in disbelief of the events as they have been outlined in this text, or as they unfold within your life, you may view the Zero Point material as something to be read and stored away for a later time; a time when you feel that the information applies.

If, for some reason, "The Shift" never completes as suggested in this text, the guidelines represented as *Awakening to Zero Point* offer powerful choices of life-affirming possibilities. It is these possibilities that we explore through choices made each day, in the presence of those that we hold most dear.

AFTERWORD

WHERE DO WE GO FROM HERE?

Awakening to Zero Point: The Collective Initiation was originally intended as a workbook supplement for day one of the three-day intensive of the same name. It was designed to provide the data, graphs and charts offering the foundation for the material that would follow. The remainder of the workshop is dedicated to the role that life relationships play toward teaching, preparing and training us to embody the ancient vibratory technology of Compassion. It is within the context of Compassion that I offer the following.

I sense that this time in our lives represents a crossroads of opportunity in our ancient memory. Twenty-five-hundred-year old-texts remind us that we are living the close of a grand cycle of experience unprecedented in human, as well as Earth history. It is now, during the closing years of this cycle, that we are invited to choose between one of two paths, each complete with its unique choices and consequences, each leading to the same destination within different time frames and experiences.

The First Path may be viewed as our existing paradigm represented as technologies engineered outside of our bodies. Expressed as extensions of ourselves interacting with the world around us, this path has been our cultural response to the challenges of life. Causes of life events are found "out there" in a world that is perceived as separate and distinct from our bodies. Therefore solutions are engineered "out there," discounting the interplay between us and our world. Illness, disease, deficiencies and conditions are viewed as originating and cured through things that we "do."

External technology is us, remembering through our machines, the same principles that we demonstrate as life in our bodies. These principles include properties of capacitance, resistance, transmission, receiving and storage of scalar and vector energies. In my years of service in the earth, space and computer sciences, I have yet to witness a technology developed outside of the human body that is not mirrored as the body itself.

External technology is us remembering ourselves by building models of ourselves, outside of ourselves, and applying the models back to ourselves. The First Path has proven itself, and its reliability to a point, representing one possibility for experience. There is another path.

The Second Path is the path of internal technology. Remembered rather than engineered, internal technology originates from within, as our expression of life. This path remembers us as the sacred union between the atomic expression of "Mother Earth" and the electrical and magnetic expression of "Father Heaven." Ancient Essene sciences emphasize this idea as the basis of their earliest teachings. Perhaps the most astounding of these references is the link between feeling, thought and their relationship to human physiology.

> *Three are the dwellings of the Son of Man, and no one may come before the face of the (One) who knows not the angel of peace in each of the three. These are body, thoughts and feelings.* (Parentheses are the author's.)
>
> ADAPTED FROM THE ESSENE GOSPEL OF PEACE[1]

As our bodies, immune systems and emotions have been taxed to unprecedented levels we are invited to make use of this power from within. The peace sought in our world, in our body and in this Essene reference is the same peace born of Compassionate emotion. Defined as a quality of thought, feeling and emotion, Compassion may be demonstrated through our choice of conduct in our daily lives.

The ancients remind us that Compassion transcends something that is "done" or accomplished. Rather, Compassion is that which you allow yourself to become. Doing is a hallmark of the paradigm that has been. We have grown beyond the doing. Put simply, to become is the very essence of the powerful and ancient message that was left to us through many texts and traditions of those who have come before us.

The message is this:

Those conditions that you most choose to have in your life, you must first become as your life.

Deceptively simple, this eloquent phrase is a summation of all of the work performed by all of the masters, all of the saints, scientists, technicians and families, each of your spiritual forbearers, preparing you for this very time in Earth and human history. Those that claim to choose peace, prosperity, health and vitality for their world must first become these very attributes themselves.

To "become" requires a shift in viewpoint. Changing your body chemistry, by shifting your viewpoint, is perhaps the single most powerful tool available to you for the remainder of your lifetime. Researchers have recently demonstrated to the Western world a phenomenon that has been taught through the mystery schools for thousands of years.

These studies:

- **Support** ancient reminders that human emotion determines the actual patterning of DNA within our bodies.[2]

It appears that the vibratory template of emotion actually "touches" the molecules of DNA in our cells, "waking up" dormant codes of immunity and vitality that may lay dormant within us. Possibly for the first time, we now have digital evidence of the technology that has been referenced, almost universally, throughout the ancient and indigenous texts. Life provides us the opportunity to experience emotion. The way that we resolve our emotion programs our DNA, affirming or denying life within our cells in response to our beliefs and attitudes.

- **Suggest** that DNA determines how patterns of light, expressed as matter, surround the human body.[3]

  Stated another way, researchers have discovered that the arrangement of matter (atoms, bacteria, viruses, climate, even other people) surrounding your body, may be directly linked to the feeling and emotion from within your body.

From these two statements alone, imagine the implications! Beyond microcircuit technology, beyond genetic splicing and drug-induced engineering, without exception this relationship between your physical body (DNA) and emotion represents the single most sophisticated technology to ever grace this world through the expression of our bodies. If it is true that emotion regulates our very genetic makeup, then why have we chosen emotions in the past that seem to deny life in our bodies? Taken a step further, why do we demand of scientists, politicians and world leaders that "they" find cures and make peace "out there"?

Perhaps the answer may be found in a recent work by Joseph Rael. In his video *The Sound Beings,* Rael[4] of the Tewa people explains our choice of medicines, vaccines and antibiotics as our way of becoming "unstuck" from a pattern of life. When we become "stuck" in a belief, thought or pattern, light cannot move through us. We experience "being stuck" as illness. Medicine is our choice, our path of engineering something "out there" to "unstick" us so that we may have motion once again.

While medicines are certainly valid, is it in our best interest to rely solely on our externally engineered solutions? Rather than engineer the technology out there to address these and other consequences of our past condition, why not accept the gift of our engineered solutions while we become the technology? Why not feel the feelings, emote the emotions and think the thoughts that allow us to shift to a state of being where bacteria, virus, change and even death are of little consequence? The cures are simply patterns of vibration engineered through our models of ourselves "out there." Why not become the models, why not become the vibrations from within?

Our own science now has demonstrated that the quality of our DNA is directly tied to our ability to forgive, allow and love through the expression of life. As our emotions choose and affirm life, our cells are bathed in the vibratory template of Compassion mirrored through our bodies. Could our greatest challenges of he relationship and survival be our way of redefining biological limits accepted long

Are we witnessing the birth of a new species of human genetically shifting to accommodate this time in history?

I invite you to explore these and other possibilities through my other work marrying science, relationships and our ancient message of Compassion. *Walking Between the Worlds: The Science of Compassion* is a reader-friendly text offering for the first time in print:

- A step-by-step guide to the ancient Science of Compassion!
- The Essene Mysteries of Relationship illustrated as case histories and true life accounts!
- The latest research confirming the role of emotion in our immune response and relationships!
- The opportunity to redefine hate, fear, light, dark, separation and the role that each plays in our lives!

Please,

use the gift of your life wisely.

NOTES TO THE READER

Note One
Validating The Shift: Is This Real?

How do you know if "The Shift" is real? How do you know whether the events portrayed in this book will come to pass? What does this information mean in terms of your day-to-day life?

Many will rest the burden of validation for *Awakening to Zero Point* upon their ability to reproduce similar information from other sources; upon a consensus from the works and teachings of others. For those individuals, I invite you to allow yourselves the freedom to move beyond the limits of your conditioning. Your teachings have been valid in a relative sense, preparing you for a much broader, all encompassing truth. Rather than viewing through the eyes of conditioning, I invite you to view your teachings as a foundation, a pivotal point from which to expand your awareness, embracing a broader knowledge. In this expansion, you may discover that the key to all knowledge returns to you. You live as a mirrored metaphor for all of the processes, all of the possibilities, all combinations of creation.

The event of The Shift may never be acknowledged in the mass media of television or syndicated publications. The events are unfolding now, and even now relatively few are able to "see" the continuity; their vision is guided by those who choose to "look" at what is pointed out to them—at what they are directed to see. You will experience The Shift uniquely as an individual. You may hear sounds, feel the vibrations and sense minute changes in the grids that those around you are not able to sense. You will experience The Shift through your own frame of reference. The concepts presented in this text are about defining that frame of reference; assuming responsibility for your health, your pain, your joy and the quality of your life experience. In knowing the mechanism of the Earth-heart-brain-immune system relationship, there is a tangible and logical link to the process of life and The Shift.

What if The Shift never happens? It is happening now—it may complete within your lifetime. If, for some reason you are never able to "see" The Shift, the tools offered through this text represent powerful options in our expression of life. The tools may have served as a catalyst, triggering your ability to remember higher options in response to life.

Note Two
Body Truth

As you read the words of this material, please bear in mind that you are receiving information on many levels in addition to the visual energy of the written form. This is intentional, as this text was developed as a tool of awakening. The resonant quality of the letter symbols on the page interact with fields of your awareness, activating and enhancing thought processes that will lift you to a heightened state of awareness. From that state, you will find that information begins to come through you rather than coming to you. You become the information.

I invite you to be less concerned with *where* this information comes from, focusing on *how* this information feels. How does each aspect of your body respond as it interacts with the thought-energy patterns generated by the word/symbols? Throughout your lifetime, you have probably come to learn through a process of comparison: a gathering of information from one source and comparing it to some other source for validation. You may ask an expert in that particular field or go to some reference body, some written material, some statistic that will lead you to believe that piece of information is valid for you. This method of validation has probably worked well for you up to a point. I encourage you to question and continue to validate information as you receive it to determine your truth.

There is, however, another form of learning. I refer to this additional method of learning as the *Resonance of Truth* or *Learning from the Heart.* If you have not already had the experience, in each lifetime we are faced with some situation, or circumstance for which we have absolutely no reference point. We face the experience alone. There will be no where to turn to, though we may be in an office building, busy street or public area. There will be no one to ask though we may be surrounded by others. *It will be an experience for which we have absolutely no point-of-reference*. Those around us will either not understand, or not be capable of the advice that we are seeking from them. All that we will have is ourselves.

This information comes to you as a reminder that *you* are all that *you* will ever need. You already have all of the tools, all of the knowledge, all of the guidance that you will ever need. All that you do is in the context of remembering, moving your awareness above the dense thoughts of Earth's gravity to that sanctuary within yourself where you exist as the blueprint of perfection. It is in this space that you will always find your truth. *Your body knows this space very well*. Each time you move into fields of information of any form, your body knows how you are relating to this energy. The information may be in the form of a spoken word. It may exist as fields of color, as when walking through a deep green forest. The information may be visual, a movie, or auditory, such as music. The subtle energetic sensors of your body will react to this information in a reliable, consistent manner that allows you to make decisions as to the suitability of this information relative to your best interests. *The sensations that your body exhibits in the presence of this information is your key to your truth.* You may refer to these sensations as your *Body Truth.*

The sensations of Body Truth may vary from person to person. They will always be constant for any one individual. Once you become aware of your body's signals to truth, you will begin to use them almost as second nature, if you do not already do so. Viewed within the context of the situation and the kind of information that you are being exposed to, body sensations provide a very good point-of-reference to your truth. Physiological responses to information may include, but not be limited to, the following:

- Localized perspiration
- Increase in body temperature
- Rapid pulse
- Tingling on the face, hands or feet
- Goose bumps
- Ringing in the ears
- Shallow respiration

For some, truth is indicated through perspiration on the brow, palms of the hands or sternum areas. This perspiration will be of a different nature than that of physical exertion, containing less salt. A common reaction is an increase in body temperature accompanied by rapid pulse and shallow respiration. Some individuals will develop the short-lived condition known as "goose bumps." Typically lasting only seconds, goose bumps are an indication of a high degree of resonance between that individual and the field of energy that they have just encountered. There are reactions too numerous to identify in this text. Ringing ears, "tingling" sensations of the face, feet and hands—all may be your body's feedback to the fields of energy/information that you have just placed yourself into. Each individual is unique and in that uniqueness, the body's reaction may vary. Become aware of how your body feels as you place yourself in various situations. You may use the reading of this book as a tool to aid in your development of your sacred gifts of "Inner Technology" and Body Truth.

GLOSSARY

Charge. A strong feeling as to the rightness, wrongness or appropriateness of an outcome is our charge upon that experience. Technically, charge may be measured as an electrical potential surrounding an expectation, act or situation. Created as an event is judged and labeled through the eyes of polarity as good, bad, dark or light, the charge is often sensed as feelings of anger, sadness or frustration. The holographic mirror of consciousness assures us that we will experience our judgments (charges) so that the judgments may be redefined (reconciled) allowing us to move forward in life without the hindrance of that particular charge.

Compassion. The single-word expression of a specific quality of thought, feeling and emotion; thought without attachment to the outcome, feeling without the distortion of an individual's life bias and emotion without the charge of polarity. The science of Compassion allows the witnessing/experience of an event without the polarized judgment as to the rightness or wrongness of the event.

Creator. One, who knowing the principles of "God," creates life from nonlife. Apart from the procreation resulting as the union of sperm and egg, creation is the act of bringing together nonliving compounds within an electrical environment, to producing living material.

Dimensionality. Dimensionality may be viewed as an attempt within the logical mind to resolve the hierarchy of energy into discrete bands, or ranges of experience, so that they may be referenced separately. The reality is that there is no separation. There are no clear boundaries between one vibrational experience and another. There is only a gradation, a phasing of varying states of experience. These altered states of awareness are known as dreams.

Though deceptively simple in appearance, this premise of "oneness" has proven difficult to resolve within the rational mind. The ancient texts however have offered this very concept, through many different languages, for thousands of years.

Emotion. Emotion is the power that you place into your thoughts to make them real. It is the scalar potential of emotion, combined with the scalar potential of thought, that produces the vector experience of reality. The waves of scalar potential produce the interference pattern of your vector reality. You experience emotion as sensation flowing, directed, or lodged within the liquid crystalline form of your body. The electrical charge that pulses through your body as life force provides the sensation of emotion. Emotion may be experienced spontaneously or as the result of a choice to be. Emotion is closely aligned with desire, the will to allow something to become so. As you truly desire to replace hate in your life with Compassion, you will sense the power flowing from your lower energy centers as a warm tingling sensation into your chest and heart centers.

Entrainment. The alignment of forces, or fields of energy, to allow maximum transfer of information or communication. For example, consider two elements adjacent to one another and each is vibrating. One is vibrating at a faster rate while the other is vibrating at a slower rate. The tendency for the element of slower vibration to synchronize and match the element of faster vibration may be considered entrainment. To the degree that the match is accomplished, we say that entrainment has occurred or the faster vibration has entrained the slower vibration.

Entropy. Entropy may be defined as a measure of the capacity of a system to undergo change. Newton's third law of thermodynamics states that at absolute zero, all molecules are perfectly aligned and motionless, with the degree of disorder (entropy) at zero. The greater the degree of disorder, the greater the entropy.

Essenes. An ancient brotherhood of unknown origin, who chose to separate themselves from the masses of their time to live the purity of traditions that had been left to them by their ancestors. Located primarily around the Dead Sea and Lake Mareotis in the 1st century, A.D., the teachings of this mysterious brotherhood have appeared in almost every country and religion, including Sumeria, Palestine, India, Tibet, China and Persia. Some Native North American tribes trace the roots of their ancestors to the clans of Essenes just after the execution of Jesus.

The ancient Essenes were communal agriculturists with no servants or slaves. They lived a structured life without meats or fermented drinks, allowing life spans of up to 120 years or more. Among well-known Essenes were the healers John the Beloved, John the Baptist, Elijah and Jesus of Nazareth.

Feeling. Feeling may be defined as the union of thought and emotion. As you experience sadness, hate, joy or Compassion, for example, you are experiencing feeling. Feeling is the sensation of emotion, coupled with the thought of what you are experiencing in the moment.

The liquid crystal resonator of the heart muscle is the focal point of feeling. It now becomes apparent why the body responds so well to love and Compassion. Through love and Compassion the heart is optimally tuned to the Earth, allowing the circuit to express fully and completely.

Examples of feeling:

When you feel love, you are feeling your thought of what the object of your love means to you coupled with the emotion of your desire.

God. The matrix of intelligence underlying all of creation. This principle provides the vibratory template upon which all of creation is "crystallized." The principle of God represents all possibilities and lives as each expression of masculine and feminine expression. From this perspective, the God Force is a living, vibratory pulse that lives in the spaces between the nothing and is inherent in all that we may experience in our world.

Grid. Reference to a two dimensional framework, providing a preferred pathway for energy-information-light to travel from point A to point B. The grid may be thought of as the substance that exists in the nothingness, the thread of fundamental intelligence that is the underlying fabric of creation.

Conceptually, a grid may be considered as an etheric network of guidelines, a meshed framework along which pulses of energy are directed. These ordered patterns of energy are typically formed of a single uniformly shaped pattern, repeating itself over and over as equally spaced expressions of the identical geometry in any two-dimensional direction. The energy of consideration may be as subtle as the microhertzian pulsations of human thought generated at hundreds of thousands of cycles per second, or as dense as Earth-resonant tones pulsating as low as 7.8 cycles per second. The energy, traveling as waves, is propagated throughout various aspects of creation along this lattice of high conductivity, transgressing the boundaries of star systems and dimensionality.

Hologram. A recursive pattern of energy (geometric, emotional, feeling, thought, consciousness or mathematical) that stands whole and complete unto itself while serving as a portion of a greater whole. For example, each cell of the human body is whole and complete unto itself, containing all of the information required to create another human body. At the same time, it is one cell of a much larger whole, the body itself.

By definition, each element of a holographic pattern mirrors all other elements of the pattern. This is the beauty of the holographic model of consciousness. Change introduced anywhere in the system is mirrored throughout the entire system.

Matrix. A collection of grids, appearing as stacked upon one another. The grids are hierarchical in nature, providing a structure for the gradual transition of energy-information from one zone of parameters to another.

The size, dimensions or lateral extent of the grid cell and overall grid *may* vary. The size and dimensions of the matrix itself *will* vary. Most matrices will be recognized as a subset of a larger grid-matrix system, and that one in turn will be a portion of an even larger parent system. For the purposes of this material, the grid-matrix relationship is a conceptual model built to understand the framework of creation and all life contained within. The matrices of creation are essentially *holographic* in nature; each cell being whole and complete unto itself, as well as part of an even larger whole. Within each single cell lives all of the information for the entire pattern to repeat itself again, and within each cell of that pattern the repetition continues.

Mer-Ka-Ba. *Mer* = Light, *Ka* = Spirit, *Ba* = Body. An Egyptian term describing the Light-Spirit-Body complex of energy that surrounds each cell of the human body individually as well as the entire body as a composite field.

Radiating electrical impulses, this particular form originates from the first eight cells immediately following conception, the cells that form the Star Tetrahedron. The *Mer-Ka-Ba* seed is formed, and remains within a point at the root or first chakra, located at the base of the spine. The pattern of these cells provides the blueprint for a radiant field of energy that extends through and beyond the physical boundaries of the body. The field appears in two dimensions as the Hebrew Star of David. The *Mer-Ka-Ba* may be thought of as a vehicle, in some references, the Light Body or Time-Space Vehicle, capable of transcending the perceived limitations of space, time, space-time, and dimensionality.

Platonic Solid. A Platonic Solid may be defined as the surfaces delineating a very special, fully enclosed volume. All lengths defining any portion of the volume are equal, as are the values of all interior angles defining the corners. Conceptually, the solid may be thought of as a single unit cell of form, repeating until it falls back upon itself with adjacent, matching unit cells. Each angle formed by the meeting of the unit cells, and the dimensions of each side of the cell are equal. At present, there are five solids known that meet this criteria. Referred to as the five regular Platonic Solids, they are illustrated below in order of increasing complexity as defined by the number of faces.

Platonic Solid	*Number of Faces*	*Edges*	*Vertices*
1. Tetrahedron	4	6	4
2. Hexahedron (Cube)	6	12	8
3. Octahedron	8	12	6
4. Dodecahedron	12	30	20
5. Icosahedron	20	30	12

All patterns of three dimensional creation, including our bodies, resolve to energy bonds resulting from one, or some combination of, these five forms.

Quanta. The vibratory patterns of light may be considered as *pulsed waves* created by discrete bursts of energy. Modern researchers recognize these bursts of energy as *quanta*, (brief, rapid pulses of light) and study this phenomenon as the science of quantum physics.

Reconcile. Within the context of *Walking Between the Worlds,* to reconcile an event or circumstance is to find a place of balance within which the event makes sense. Reconciliation of an event does not indicate a condoning or approval of what has happened. It simply allows for an acknowledgment within the individual so that they may move forward in life. The act of blessing, acknowledging the divine nature of life's offering, is an example of reconciliation.

Resolve. The removal of a charge upon an event, circumstance or situation is the zeroing out of the electrical potential that judgment has placed upon the event. We say that the event has been "resolved." To resolve the charge upon an event, for example, is to redefine the meaning of that event, refining the meaning until a true neutral feeling is reached. Neutrality is a biochemical expression of resolution.

Resonance. An exchange of energy between two or more systems of energy. The exchange is two-way, allowing each system to become a point of reference for the other. A common example of resonance is illustrated with two stringed instruments placed on opposite sides of the same room. As the lowest string of one instrument is plucked the same string on the second instrument will vibrate. No one touched the string, it is responding to the waves of energy that traveled across the room and found resonance with the second string. In my text, I speak of resonance between systems of energy such as the Earth and the human heart, or two individuals "tuning" to resonance through emotion.

Scalar Potential. A quality of energy described as having not been dispersed or dissipated. Scalar energy may be thought of as energy that is fully enabled and waiting to be used; a potential force available for activation. Upon activation, the potential becomes "real," or a vector quantity, that may be measured as magnitude and direction.

Shift of the Ages. Both a time in Earth history and an experience of human consciousness. Defined by the convergence of decreasing planetary magnetics and increasing planetary frequency upon a point in time, The Shift of the Ages, or simply The Shift, represents a rare opportunity of collectively repatterning the expression of human consciousness.

The Shift is the term applied to the process of Earth accelerating through a course of evolutionary change, with the human species linked, by choice, to the electromagnetic fields of Earth, following suit through a process of cellular change.

The One. A nonreligious term in reference to the matrix of intelligence underlying all of creation. This principle provides the vibratory template upon which all of creation is "crystallized." The principle of The One represents all possibilities and lives as each expression of masculine and feminine expression. From this perspective, The One is a living, vibratory pulse that lives in the spaces between the nothing and is inherent in all that we may experience in our world.

Thought. Thought may be considered as an energy of scalar potential, the directional seed of an expression of energy that may, or may not, materialize as a real or vector event. A virtual assembling of your experience, thought provides the guidance system, the direction, for where the energy of your attention may be directed. Without the influx of emotional energy into your thoughts, they are impotent and powerless to create. In the absence of power, your thought may be considered as a model or simulation of how or where your power may be directed. Fantasies, "what ifs," affirmations, and "I choose to," are examples of the beginning of a thought. These processes will determine where you focus your attention.

Vesica Piscis. As two perfect spheres are formed overlapping one another by half (each contains half the diameter of the other), a zone of commonality is created in the overlap. In the science of Sacred Geometry, this zone is referred to as the *Vesica Piscis*. The form of the *vesica* was used as both the Egyptian glyph for the "mouth" as well as for the Creator. Additionally, this glyph is also very similar to the Mayan glyph for zero, associated with our galaxy of the Milky Way.

Zero Point. The amount of vibrational energy associated with matter, as the parameters defining that matter decline to zero. To an *observer,* the world at Zero Point appears to be very still, while the *participant* experiences a quantum restructuring of the very boundaries that define the experience. Earth and our bodies are preparing for the Zero Point experience of change, collectively known by the ancients as The Shift of the Ages.

REFERENCES

INTRODUCTION

1 William Grieder, "The Secret Plan to Bail Out Banks," *San Jose Mercury News*, February 16, 1992

2 Richard Burke, "Bankruptcies Still Setting Record Pace," *Knight-Ridder Newspapers*, February, 1992

3 John Anthony West, personal conversation, Author of *Serpent in the Sky*, Cairo, Egypt, November, 1992

4 Richard Hoagland, *Hoagland's Mars: Vol.II, The U.N. Briefing, The Terrestrial Connection* (video), B.C. Video Inc., New York City, 1992

5 Elisha Qimron, "Paying the Price For Freeing the Scrolls," *Biblical Archaeology Review*, Vol. 19, Number 4, July/August, 1993

6 Robert S. Boyd, "Weather Went to Extremes in '93," *San Jose Mercury News*, January 12, 1994

7 Michael D. Lemonick, "Ozone Breakdown," *Time Magazine*, February, 17, 1992

8 Christine Gorman, "Invincible Aids," *Time Magazine*, August 3, 1992

9 Lee Siegel, "More Evidence That The Big One Is Overdue," *Associated Press*, January, 1993

10 Christine Gorman, "Invincible Aids," *Time Magazine*, August 3, 1992

11 Sharon Begley, "The End of Antibiotics," *Newsweek*, March 28, 1994

CHAPTER 1
THE SHIFT OF THE AGES

1 "Body Mind Spirit," *Native American Prophecies: Tales of the End Times*, January/February, 1993

2 Frank Waters, *Mexico Mystique*, Ohio University Press, 1989

3 Jose Arguelles, *The Mayan Factor*, Bear & Company, Santa Fe, New Mexico, 1987

4 Enoch, *The Book of Enoch the Prophet*, translated from an Ethiopic manuscript in the Bodleian Library by the late Richard Laurence, LL.D., Archbishop of Cashel, Wizards Bookshelf, San Diego, California, 1983

5 Ibid.

6 *The Holy Bible*, Revised Standard Version, Concordance, The World Publishing Company, 2231 West 110 Street, Cleveland, Ohio, 1962

7 Ibid.

8 Thoth, *The Emerald Tablets of Thoth*, Translation by Doreal, Source Books, Mt. Juliette, Tennessee, 1996

9 Ibid.

10 L. Don Leet, Sheldon Judson, *Physical Geology*, Prentice-Hall, Inc., 1971

11 Nils-Axel Morner, "Earth's Rotation and Magnetism," Article from *New Approaches in Geomagnetism and the Earth's Rotation,* Stig Floodmark, University of Stockholm, Sweden, 1988

12 L. Don Leet, Sheldon Judson, *Physical Geology*, Prentice-Hall, Inc., Englewood Cliffs, New Jersey, 1971

13 Nils-Axel Morner, "Earth's Rotation and Magnetism," Article from *New Approaches in Geomagnetism and the Earth's Rotation,* Stig Floodmark, University of Stockholm, Sweden, 1988

14 Zecharia Sitchin, *The Lost Realms*, Avon Books, 1990

15 *The Holy Bible*, Revised Standard Version, Concordance, The World Publishing Company, 2231 West 110 Street, Cleveland, Ohio, 1962

16 Richard Monastersky, "The Flap Over Magnetic Flips," *Science News*, Vol. 14, June 12, 1993

17 L. Don Leet, Sheldon Judson, *Physical Geology*, Prentice-Hall, Inc., Englewood Cliffs, New Jersey, 1971

18 Tsuneji Rikitake and Yoshimori Honkura, *Solid Earth Geomagnetism,* Terra Scientific Publishing Co., Tokyo, Japan, 1985

19 Nils-Axel Morner, "Earth's Rotation and Magnetism," Article from *New Approaches in Geomagnetism and the Earth's Rotation,* Stig Floodmark, University of Stockholm, Sweden, 1988

20 R. Cowen, "Sound Waves May Drive Cosmic Structure," *Science News*, January 11, 1997

21 Bob Brown, *Propagation,* WorldRadio, March, 1995

22 L. Don Leet, Sheldon Judson, *Physical Geology*, Prentice-Hall, Inc., Englewood Cliffs, New Jersey, 1971

23 Ibid.

24 Frank Waters, *Book of the Hopi,* Ballentine Books, Inc. New York, 1963

25 Jose Arguelles, *The Mayan Factor*, Bear & Company, Santa Fe, N.M., 1987

26 Joseph Smith, *The Book of Mormon*, The Church of Jesus Christ of Latter-day Saints, Salt Lake City, Utah, 1961

CHAPTER 2
THE LANGUAGE OF CREATION

1 F. Donald Bloss, *Crystallography and Crystal Chemistry*, Holt, Rinehart and Winston, Inc., New York, 1971

2 Eugene Mallove, "The Cosmos and the Computer: Simulating the Universe," *Computers in Science*, Vol. 1, No. 2, 1987

3 Tom Kenyon, Founder and President, *Acoustic Brain Research,* P.O. Box 16427 Chapel Hill, North Carolina, 27516, 1996

4 C.W. Snell, *Sympathetic Vibratory Physics*, The Snell Manuscript, Edited by Dale Pond, Delta Spectrum Research, Inc., Colorado Springs, Colorado, 1986

5 Alex Gray, *Sacred Mirrors, The Visionary Art of Alex Gray,* Inner Traditions International, Rochester, Vermont, 1990

6 Ibid.

7 Thoth, *The Emerald Tablets of Thoth*, Translation by Doreal, Source Books, Mt. Juliette, Tennessee, 1996

8 Paramahansa Yogananda, *Scientific Healing Affirmations*, Self-Realization Fellowship, Ninth Printing, 1990

9 Robert Lawlor, *Sacred Geometry, Philosophy and Practice*, Thames and Hudson Ltd., London, 1982

10 Hans Jenny, *Cymatics,* Basilius Presse AG, Switzerland, 1974

11 Thoth, *The Emerald Tablets of Thoth*, Translation by Doreal, Source Books, Mt. Juliette, Tennessee, 1996

12 Robert Lawlor, *Sacred Geometry, Philosophy and Practice*, Thames and Hudson Ltd., London, 1982

13 Cornelius S. Hurlbut, Jr., Cornelius Klein, After J.D. Dana, *Manual of Mineralogy*, John Wiley and Sons, New York, 1971

14 Thoth, *The Emerald Tablets of Thoth*, Translation by Doreal, Source Books, Mt. Juliette, Tennessee, 1996

15 Drunvalo Melchizedek, *The Flower of Life Workshop and Meditation,* Hummingbird, Prescott, Arizona, 1996

16 Thoth, *The Emerald Tablets of Thoth*, Translation by Doreal, Source Books, Mt. Juliette, Tennessee, 1996

17 Ibid.

18 John Davidson, *The Secret of the Creative Vacuum,* The C.W. Daniel Company, Limited, 1989

19 Alexander Joseph, *The Brilliant Eye*, 151 Concord Bridge, Big Water, Utah, 84741, 1991

20 Drunvalo Melchizedek, *The Flower of Life Workshop and Meditation,* Hummingbird, Prescott, Arizona, 1996

21 Thoth, *The Emerald Tablets of Thoth*, Translation by Doreal, Source Books, Mt. Juliette, Tennessee, 1996

CHAPTER 3
INTERDIMENSIONAL CIRCUITRY

1 Chris Griscom, *Ageless Body*, Light Institute Press, 1992

2 James D. Watson, *Molecular Biology of the Gene*, W.A. Benjamin, Inc. 1976

CHAPTER 4
CROP CIRCLES

1 Colin Andrews, *Undeniable Evidence*, Colin Andrews, video produced by Ark Soundwaves of Glastonbury, Ltd., 1991

2 Ibid.

3 Ibid.

4 Hans Jenny, *Cymatics, Bringing Matter to Life With Sound*, video produced by MACROmedia, Brookline, Massachusetts, 1986

5 Colin Andrews, *Undeniable Evidence*, Colin Andrews, video produced by Ark Soundwaves of Glastonbury, Ltd., 1991

6 Beth Davis, *Ciphers in the Crops*, Gateway Books, The Hollies, Wellow, Bath, BAA2 8QJ, United Kingdom, 1991

7 Norman Rothwell, *Understanding Genetics*, Oxford University Press, 1988

8 Thoth, *The Emerald Tablets of Thoth*, Translation by Doreal, Source Books, Mt. Juliette, Tennessee, 1996

9 Jesse Walter Fewkes, *Designs on Prehistoric Hopi Pottery*, Dover Publications, New York, 1973

10 R. H. Warring, *Understanding Electronics,* 2nd Edition, printed by permission of Lutterworth Press by Tab Books, 1978

11 Barbara Hand-Clow, *The Age of Light: Decoding the Photon Band—1962 to 2012*, Bear & Company, Santa Fe, New Mexico, 1993

CHAPTER 5
MESSAGE OF ZERO POINT

1 Joseph Smith, *The Book of Mormon*, The Church of Jesus Christ of Latter-day Saints, Salt Lake City, Utah, 1961

2 *The Lost Books of the Bible* and *The Forgotten Books of Eden*, World Publishing Company, New York, 1963

3 *In Search of Historic Jesus*, Sun Classics Video, Tulsa, Oklahoma, 1979

4 Robert Boissiere, *Meditations With the Hopi,* Bear & Company, Santa Fe, New Mexico, 1986

5 Elisha Qimron, "Paying the Price For Freeing the Scrolls," *Biblical Archaeology Review*, Vol. 19, Number 4, July/August, 1993

6 *The American Heritage Dictionary of the English Language*, William Morris, Editor, American Heritage Publishing Co., Inc. and Houghton Mifflin Company, New York, 1971

7 Alan Cohen, *The Peace That You Seek*, Alan Cohen Publications, Somerset, New Jersey, 1992

8 *The Essene Gospel of Peace*, Book Four, Compared, Edited and Translated by Edmond Bordeaux Szekely, Third Century Aramaic Manuscript and Old Slavic Texts, I.B.S. Internacional, Matsqui, B.C., Canada, 1937

9 Kahlil Gibran, *The Prophet,* Alfred A. Knopf, Inc. New York, 1977

10 John Davidson, *The Secret of the Creative Vacuum*, The C.W. Daniel Company Limited, 1989

11 *The Holy Bible*, Revised Standard Version, Concordance, The World Publishing Company, 2231 West 110 Street, Cleveland, Ohio, 1962

CHAPTER 6
FROM SEPARATION TO UNION

1 *Meetings with Remarkable Men*, Gurdjieff's Search for Hidden Knowledge, Corinth Video, 1987

AFTERWORD

1 *The Essene Gospel of Peace*, Book Four, Compared, Edited and Translated by Edmond Bordeaux Szekely, Third Century Aramaic Manuscript and Old Slavic Texts, I.B.S. Internacional, Matsqui, B.C., Canada, 1937

2 Dan Winter, *Alphabet of the Heart: The Genesis in Principle of Language and Feeling*, Waynesville, North Carolina

3 Vladimir Poponin, *The DNA Phantom Effect: Direct Measurement of a New Field in the Vacuum Substructure*, Institute of HeartMath, Boulder Creek, California

4 Joseph Rael, *The Sound Beings*, Exclusive Pictures/Heaven Fire Productions, Video, 1995, Van Nuys, California

ADDITIONAL BOOK REFERENCED

The Sacred Landscape, Fredrick Lehrman, Celestial Arts Publishing, P.O. Box 7327, Berkeley, California, 94707

BIBLIOGRAPHY

The American Heritage Dictionary of the English Language, (New York, New York) 1971

Andrews, Colin. *Undeniable Evidence*, (Glastonbury, England) 1991

Arguelles, Jose. *The Mayan Factor*, (Santa Fe, New Mexico) 1987

Begley, Sharon. "The End of Antibiotics," *News Week*, 1994

Bloss, Donald, F. *Crystallography and Crystal Chemistry*, (New York) 1971

"Body Mind Spirit," *Native American Prophecies: Tales of the End Times*, 1993

Boissiere, Robert. *Meditations With the Hopi*, (Santa Fe, New Mexico) 1986

Boyd, Robert S. "Weather Went to Extremes in '93," *San Jose Mercury News*, (San Jose, California) 1994

Brown, Bob. *Propagation*, WorldRadio, March, 1995

Burke, Richard. "Bankruptcies Still Setting Record Pace," *Knight-Ridder Newspapers*, 1992

Cheney, Margaret. *Tesla: A Man Out of Time*, 1981

Cohen, Alan. *The Peace That You Seek*, (Somerset, New Jersey) 1992

Cowen, R. "Sound Waves May Drive Cosmic Structure," *Science News*, (Marion, Ohio) 1997

Davidson, John. *The Secret of the Creative Vacuum*, 1989

Davis, Beth. *Ciphers in the Crops*, (The Hollies, Wellow, Bath, United Kingdom) 1991

Enoch. *The Book of Enoch the Prophet*, (San Diego) 1983

Fewkes, Jesse, Walter. *Designs on Prehistoric Hopi Pottery*, (New York, New York) 1973

Gibran, Kahlil. *The Prophet*, (New York, New York) 1977

Gorman, Christine. "Invincible Aids," *Time Magazine*, 1992

Gray, Alex. *Sacred Mirrors, The Visionary Art of Alex Gray*, (Rochester, Vermont) 1990

Grieder, William. "The Secret Plan to Bail Out Banks," *San Jose Mercury News*, (San Jose, California) 1992

Griscom, Chris. *Ageless Body*, (Galisteo, New Mexico) 1992

Hand-Clow, Barbara. *The Age of Light: Decoding the Photon Band—1962 to 2012*, (Santa Fe, New Mexico) 1993

Hoagland, Richard. *Hoagland's Mars: Vol.II, The U.N. Briefing, The Terrestrial Connection* (New York, New York) 1992

Holy Bible. Revised Standard Version, Concordance, (Cleveland, Ohio) 1962

Hurlbut, Cornelius, S. Jr. and Klein, Cornelius. *Manual of Mineralogy*, 19th Edition, (New York, New York) 1971

In Search of Historic Jesus, Sun Classics Video, (Tulsa, Oklahoma) 1979

Jenny, Hans. *Bringing Matter to Life With Sound*, Video, (Epping, New Hampshire) 1986

Ibid, *Cymatics*, Basilius Presse AG, Switzerland, 1974

Joseph, Alexander. *The Brilliant Eye*, (Big Water, Utah) 1991

Kenyon, Tom. founder of *Acoustic Brain Research*, (Chapel Hill, North Carolina) 1996

Lawlor, Robert. *Sacred Geometry, Philosophy and Practice*, (London, England) 1982

Lehrman, Fredrick. *The Sacred Landscape*, (Berkeley, California) 1988

Lemonick, Michael, D. "Ozone Breakdown," *Time Magazine,* 1992

Leet, Don, L., Judson, Sheldon. *Physical Geology*, Prentice-Hall, Inc., 1971

Mallove, Eugene. "The Cosmos and the Computer: Simulating the Universe," *Computers in Science*, 1987

Meetings with Remarkable Men, Gurdjieff's Search for Hidden Knowledge, Corinth Video, 1987

Melchizedek, Drunvalo. *The Flower of Life Workshop and Meditation,* (Prescott, Arizona) 1996

Monastersky, Richard. "The Flap Over Magnetic Flips," *Science News*, (Marion, Ohio) 1993

Morner, Nils-Axel. *Earth's Rotation and Magnetism,* (Sweden) 1982

Poponin, Vladimir. *The DNA Phantom Effect: Direct Measurement of a New Field in the Vacuum Substructure*, Institute of HeartMath, (Boulder Creek, California) 1995

Qimron, Elisha. "Paying the Price For Freeing the Scrolls," *Biblical Archaeology Review*, 1993

Rael, Joseph. *The Sound Beings*, Exclusive Pictures/Heaven Fire Productions, Video, (Van Nuys, California) 1995

Rikitake, Tsuneji, and Honkura,Yoshimori. *Solid Earth Geomagnetism,* (Tokyo, Japan) 1985

Rothwell, Norman. *Understanding Genetics*, (Oxford, England) 1988

Runkorn, S.K. Editor, *Paleogeophysics*, (Newcastle Upon Tyne, England) 1970

Siegel, Lee. *More Evidence That The Big One Is Overdue*, (Associated Press) 1993

Sitchin, Zecharia. *The Lost Realms*, 1990

Smith, Joseph. *The Book of Mormon*, (Salt Lake City, Utah) 1961

Snell, C.W. *Sympathetic Vibratory Physics,* (Colorado Springs, Colorado) 1986

Szelekey, Edmond, Bordeaux. *The Essene Gospel of Peace,* Book Four, (Matsqui, B.C., Canada), 1937

The Lost Books of the Bible and *The Forgotten Books of Eden*, World Publishing Company, (New York, New York) 1963

Thoth. *The Emerald Tablets of Thoth*, Translation by Doreal, (Mt. Juliette, Tennessee) 1994

R. H. Warring, R. H. *Understanding Electronics,* (Blue Ridge Summit, Pennsylvania) 1978

Waters, Frank. *Mexico Mystique*, (Ohio) 1989

Watson, James, D. *Molecular Biology of the Gene,* 1976

West, John, Anthony. personal conversation, (Cairo, Egypt) 1992

Winter, Dan. *Alphabet of the Heart: The Genesis in Principle of Language and Feeling*, (Waynesville, North Carolina) 1994

Yogananda, Paramahansa. *Scientific Healing Affirmations*, 1990

About the Author

Author, lecturer and guide to sacred sites throughout the world, Gregg has been featured on radio and television programs nationwide. Following his books *Awakening to Zero Point: The Collective Initiation* and *Walking Between the Worlds*, he has been a popular guest and keynote speaker for conferences, expos, and media specials regarding ancient wisdom and planetary change.

Professional careers as an earth scientist and aerospace engineer have provided Gregg with the tools to offer his powerful seminars with clarity and relevance. Two near-death experiences early in life provide the intimate language to express his message of hope and opportunity.

Gregg provides his powerful message of Compassion as a series of one-, two- and three-day seminars. Each workshop is a multimedia experience of sight, sound, feeling and ceremony as you are skillfully guided through your memory of the vibratory technologies of Compassion, emotion and relationships.

The solitude of northern New Mexico's mountains and South Florida's coasts serve as home and inspiration to Gregg and his wife Melissa between their travels. Additionally, as the timing and conditions allow, they lead journeys to sacred sites in Egypt, Peru, Bolivia, Tibet and the American Desert Southwest. For more information regarding Gregg's workshops, seminars and Sacred Journeys, please contact his office at:

1-500-675-6308

You may write to Gregg Braden at the following address.

c/o Sacred Spaces/Ancient Wisdom
H.C. 81 Box 683
Questa, New Mexico 87556
Attn: Personal

Schedules

Gregg's extensive lecture and travel itinerary has necessitated the development of an office staff to assist him with the volumes of daily correspondence. Your questions and comments are important. He and his staff work closely to honor each request and inquiry. Thank you in advance for your patience. For dates, locations and specifics of seminars and workshops please address your correspondence to:

Sacred Spaces/Ancient Wisdom
H.C. 81 Box 683
Questa, N.M. 87556
Attn: Schedules

1-500-675-6308
Outside the U.S. 1-505-424-6892

Books, Video and Audio Recordings

To order additional copies of this book and related material, please use the following address:

Radio Bookstore Press
P.O. Box 3010
Bellevue, Wa. 98009-3010

To order books tapes and videos by telephone,
please call 1-800-243-1438.
Outside the U.S. and in Seattle, Wa. 1-425-455-1053

Walking Between the Worlds: The Science of Compassion

by Gregg Braden

Ancient calendars indicate that we are living the completion of a grand cycle of human experience. Within the last years of this cycle, we have been asked to accommodate greater change in less time than at any other point in recorded human history. Our bodies, immune systems and emotions have been challenged to unprecedented levels. At the same time, science is witnessing phenomenon for which there are no reference points of comparison. Two thousand year old texts remind us that compassion is an accessible state of awareness determining the quality of our well being. Are we witnessing the birth of a new species of human genetically shifting to accommodate this time of change? Recent data demonstrates that compassionate emotion may be our forgotten switch to turn "ON" powerful codes of genetic options. Gregg Braden explores these and other possibilities in his latest work marrying science, relationships and the ancient messages of compassion.

#5589 Book - Softcover - 248 pages - List ~~$17.95~~ **$14.99** SAVE 17%

The Isaiah Effect: Decoding the Lost Science of Prayer & Prophecy

by Gregg Braden

Of all the Dead Sea Scrolls, Gregg finds the Isaiah Scroll of particular interest. Gregg examines aspects of the Isaiah prophecies, ancient Hopi, Mayan, and Egyptian texts, laboratory studies on the power of prayer, and emerging ideas from quantum physics. He finds clues to the mystical power of "active prayer" -- a lost art that is key to choosing our future. Gregg calls this the "Isaiah Effect."

#5587 Book - Softcover - 276 pages - List ~~$14.00~~ **$12.69** SAVE 10%

Awakening to Zero Point: The Video

Presented by Gregg Braden

This fully-illustrated, multi-media presentation is a message of hope and compassion. Gregg presents evidence that we are collectively preparing for an evolutionary leap. He draws from the records of ancient cultures, new scientific research on the connection between our emotions and DNA, and geological data on Earth's cycles. One cycle of great interest is the strength of Earth's magnetic field; science does not yet acknowledge the subtle but profound effect this and other cyclic conditions have on us. Previous cultures have either experienced these cycles, or similated them at sacred sites and temples, leaving us a "road map." As these cycles converge now, we are allowed graceful access to higher states of consciousness. Gregg details: How to change the patterns that can determine how and why you love, fear, judge, feel, need, and hurt. The cause for dramatic shifts you may be experiencing (radical changes in sleep patterns and dream states, perception of time "speeding up" and intensified emotions and relationships). The direct link between what you feel and how your body codes genetic information. How to affect positive enhancement of your own physiology.

#7401 2-VHS Videos, 3 hrs 40 min - List ~~$39.95~~ **$35.99** SAVE 10%

#7405 2-VHS/PAL, European Format - List ~~$49.95~~ **$44.99** SAVE 10%

Walking Between The Worlds: The Audio Book

by Gregg Braden

A five and a half hour audio presentation by Gregg Braden, based upon the book and seminar ***Walking Between The Worlds.*** Gregg finds evidence to support ancient texts that speak of compassion as a "path" allowing a life of vitality, longevity, and freedom from disease, in new scientific studies that suggest direct links between the emotion of compassion, our DNA, and the immune system. Gregg revives the "Science of Compassion" in a sequence of emotional, logical, and vibratory "codes" and asks: Is Emotion our forgotten "switch" that can turn on our genetic codes at will? What are the verbal "equations" for compassion, left to us over 2,000 years ago? Do we have a "seventh sense," one that links our experience with the way that we "feel" about our experience?

#6113 6 - audio cassettes 5 1/2 hrs - List ~~$39.95~~ **$35.99** SAVE 10%

Walking Between the Worlds: The Video

Presented by Gregg Braden

Gregg Braden further explores the understanding the inner technology of emotions by heavily detailing the Essene Mirrors of Relationship. The ancient knowledge of compassion may be one of the few viable approaches for creating a better environment for self-knowledge - now and in the future.

#7402 2 VHS Videos - 4 hrs - List ~~$39.95~~ **$35.99** SAVE 10%

Beyond Zero Point

Gregg Braden

Gregg offers scientific research essential to understanding the prophecies of the Hopi, Essenes, Mayans, Egyptians, and others. Experience how the outer world mirrors the inner conditions of our consciousness and transcends ordinary awareness.

#6114 Audio Cassettes - 2 - List ~~$18.95~~ **$17.09** SAVE 10%

The Lost Mode of Prayer

Gregg Braden

Gregg delves into libraries lost to the West nearly a millennia ago. The Essenes were but a single link in a secret chain of ancient wisdom holders. Gregg finds in their records, an esoteric practice of focusing thoughts and emotions to directly influence the physical world and those around us.

#6115 Audio Cassettes - 2 - List ~~$18.95~~ **$17.09** SAVE 10%

Audio Isaiah Effect (Abridged)

read by Gregg Braden

Seventeen hundred years ago, key elements of our ancient heritage were lost, relegated to the esoteric traditions of mystery schools and sacred orders. Among the most empowering of the forgotten elements are references to a science with the power to bring everlasting healing to our bodies and initiate an unprecedented era of peace and cooperation between governments and nations.

#6116 Audio Cassettes - 2 - List ~~$19.95~~ **$17.99** SAVE 10%

INTERVIEWS ON AUDIO CASSETTE

The Collective Initiation

Gregg Braden

Highlights: Humankind is fast approaching a collective evolutionary leap. Sacred sites and ancient texts suggest that previous cultures constructed temples which simulated the geophysical parameters that hasten higher consciousness. Evidence that these same cycles (Earth's diminishing magnetic field and rising base frequency) are converging today.

#1122 Interview on Cassettes - 2 - List ~~$14.95~~ **$13.49** SAVE 10%

Decoding Crop Circles

Gregg Braden

Highlights: Gregg views crop circle designs as glyphs, symbols meaningful in the languages of genetics, physics, & sacred geometry. He offers an interpretation for the message of the circles.

#1259 Interview on Cassettes - 2 - List ~~$14.95~~ **$13.49** SAVE 10%

The Physics of Emotion

Gregg Braden

Highlights: Emotions may serve as a "switch," turning on or off specific DNA codons. Why our lives are the continuation of spiritual initiations begun thousands of years ago.

#1357 Interview on Cassettes - 2 - List ~~$14.95~~ **$13.49** SAVE 10%

Tools for Inner Technology

Gregg Braden

Highlights: Many cultures had specific insights into emotion as a means to accelerate spiritual growth. Gregg details "the Science of Compassion" of the ancient Essenes.

#1437 Interview on Cassette - 1 - List ~~$7.95~~ **$7.19** SAVE 10%

Technology to Tune Emotions

Gregg Braden

Highlights: New technology may be imitating a lost ancient technology and therapy. Gregg looks at heiroglyphs and other clues that a lost technology was used to tune emotions to specific frequencies, with specific postive effect.

#1676 Interview on Cassette - 1 - List ~~$7.95~~ **$7.19** SAVE 10%

Earth Magnetics and Ancient Wisdom

Gregg Braden

Highlights: Anomalies in Earth's magnetic field demonstrates the wandering of magnetic North. Airport ground control recalibrates instruments to align with North. Gregg cites references from previous cultures on how they saw similar anomalies unfolding long ago.

#1684 Interview on Cassettes - 2 - List ~~$14.95~~ **$13.49** SAVE 10%

DNA Transmutations

Gregg Braden

Highlights: Gregg discusses the latest findings in genetic science that confirms our DNA can and does change according to specific conditions in the environment, and in response to emotion. Positive and negative emotions act very differently upon our DNA. We can choose how we respond, emotionally, to the events around us, and we can activate positive emotions intentionally. The effect on our physiology when we do.

#1938 Interview on Cassette - 1 - List ~~$7.95~~ **$7.19** SAVE 10%

The Power of Prayer

Gregg Braden

Highlights: Gregg looks for confirmation of wisdom expressed in ancient texts from the Middle East, Tibet, Peru, and the American Southwest in laboratory experiments on particle physics, remote viewing, and the power of prayer.

#2001 Interview on Cassette - 1 - List ~~$7.95~~ **$7.19** SAVE 10%

Revival of Lost Spiritual Techniques

Gregg Braden

Through a language that we are just beginning to understand, we are reminded of two empowering technologies that allow us to determine the conditions of our bodies, and the future of our world. Join Gregg as he addresses these topics, updates us on his recent journeys to Egypt and Tibet, and shares his latest findings on the "lost" Dead Sea Scroll that unexpectedly resurfaced in late 1999.

#2004 Interview on Cassette - 1 - List ~~$7.95~~ **$7.19** SAVE 10%

The Emotion is the Power

Gregg Braden

Highlights: It's all in the feeling, the emotion, explains Gregg, and ancient cultures from the Essenes to the Tibetans to the Inca knew this. He details how we today can incorporate this understanding of the art of manifesting and the power of prayer into choosing our future among all the possible futures lying dormant.

#2033 Interview on Cassette - 1 - List ~~$7.95~~ **$7.19** SAVE 10%

Experimental Evidence for the Quantum Hologram

Gregg Braden

Highlights: Gregg outlines three recent experiments that indicate the presence of a field that binds all of creation and operates with the laws of Quantum physics laws. He also sheds new light on ancient textual references to this field, and how our emotions and state of mind influence our DNA, which in turn influences this field.

#2163 Interview on Cassette - 1 - List ~~$7.95~~ **$7.19** SAVE 10%

VIDEO PRESENTATIONS

Sacred Living Geometry

by Callum Coats

Illustrated lecture with Callum Coats, who spent 20 years researching Austrian ecologist Viktor Schauberger. A fascinating and in-depth look at Schauberger's ahead-of-his-time environmental theories, observations of Nature, and technical devices, which hold immense promise for the future.

#7590 2-VHS Videos - 3 hrs - List ~~$39.95~~ **$34.95** SAVE 13%

Dowsing for Spiritual Transformation

by Eric Dowsett

Australian author and teacher Eric Dowsett's use of dowsing as a spiritual practice has taken dowsing to the next level, incorporating true transformation into the process. His teaching of simple yet powerful techniques to locate and clear disturbances in the environment, self, and others is a revolutionary next step in the world of energy work. With the exploding interest in dowsing as a tool for practical answers, his work provides information people can put to immediate use in their lives, as well as provides an experiential journey with compassion to an open heart.

#7815 VHS Video - 2 hrs - List ~~$29.95~~ **$24.95** SAVE 13%

Secrets of the Bird Tribe: Lost Stargate Artifacts & Spiritual Teachings

by William Henry

An illustrated lecture presentation. William Henry interprets the symbol of the bird man as ancient winged extraterrestrial beings and their stargate artifacts, and stargate myths of Moses, Nebuchadnezzar and other historical figures in line with the An-nun-aki of Zecharia Sitchin's research. Henry offers startling new insights into the ancient secrets for cleansing the human soul, purifying the human heart and the means to increase our spritual vision to perceive the gateways which prophecies predict will soon be opening.

#7780 VHS Video - 1 hr 45min - List ~~$24.95~~ **$19.99** SAVE 20%

The Science of Innate Intelligence: The Role of Belief and Thought Mechanisms in Health

by Bruce Lipton

In this illustrated lecture, Bruce Lipton demonstrates how our thoughts and perceptions actively affect our health and behavior. Discover new information about the biology of consciousness, including its actual molecular mechanisms. Explore how knowledge of consciousness mechanisms can be employed to redefine our physical and emotional well being. Learn how our perceptions have been directly linked to the production of gene-altering mutations. This two part video presentation, featuring captivating, scientific, well referenced material, has been approved for Continuing Education Credit for health care providers.

#7431 2-VHS Video - 2 hrs 30 min - List ~~$59.95~~ **$53.99** SAVE 10%

The Code - Video

Carl Munck

An illustrated lecture presentation, this homespun video was produced by Munck. Decoding a mathematically based key to ancient temples, from Stonehenge to the pyramids of Egypt and Mexico. Using illustrations and simple math equations, Munck describes how pyramids and earthen mounds around the world "know where they are."

#7610 VHS Video - 2 hrs - List ~~$24.95~~ **$19.99** SAVE 20%

The Code: Parts 2 & 3

Carl Munck

Archeocryptography school is again in session as *The Code: Parts 2 & 3* continue here on one video with Carl Munck as your guide and teacher. Geomathematical logic, as well as contemporary science, demonstrates that the ancients measured and mapped the Earth. Part 2 - Matrix West: structures in N. & S. America Part 3 - The Pi Pyramids: the relationships of many structures and Pi.

#7611 VHS Video - 1 hr 45 min - List ~~$29.95~~ **$24.99** SAVE 17%

The Code 2000

Carl Munck

Carl tries to "avoid the math" as he clearly demonstrates the simple, yet beautiful elegance with which ancient monuments and pyramids were placed on the global matrix to let anyone know why there are where they are. Learn how the "Temple of the Giant Jaguar" at Tikal points to sunken Lemurian ruins off the coast of Okinawa, (by using the sunken pyramids at Rock Lake as a clue,) that were recently discovered by a scuba team. Now, learn how these Lemurian ruins connect themselves to Tiahuanaco and the Pyramids of China's Xian Province; again, without using Code Math. Learn the relationship between Olympus Mons, Jupiter's Great Red Spot and the *former* position of The Great Pyramid and how they relate to Hyperdimensionality and Anti-Gravity. Much more is covered in Carl's most authoritative look yet at the global matrix. A must have for Code buffs.

#7612 VHS Video - 2 hrs - List ~~$29.95~~ **$24.99** SAVE 17%

American Discovery - the Video

Gunnar Thompson

Gunnar Thompson Ph.D takes us on a journey into the past to find the key to the future. Join us for a tour of America's hidden heritage of multicultural discovery. Who were the voyagers from all over the world who reached the Americas long before Columbus? What artifacts and other evidence were left behind by hundreds of years of early contact and trade? What ancient, closely-held maps, revealing coastline sections of North America, guided Columbus on his own voyage? Where are these maps today? What are the implications of reclaiming America's true multicultural history? Gunnar has followed the clues to New World Discovery on his travels around the world. He is the leading scholar on the subject of multicultural voyages to the New World before Columbus.

#7810 VHS Video - 1 hr 45 min - List ~~$29.95~~ **$24.99** SAVE 17%

The Hidden History of the Human Race: The Video

Produced and Directed by Richard Thompson

Computer generated illustrations and graphics, with commentary by Richard Thompson, co-author of the books *Forbidden Archeology* and *The Hidden History of the Human Race*. He presents a controversial challenge to rethink our understanding of human origins, identity and destiny: How science suppresses anomalies that run counter to the current favored theories. Why the scientific method dictates exploration of all theories suggested by all the evidence. One theory, suggested by almost half the artifacts we have so far of human antiquity, is that humans like us have lived on Earth for millions of years. This is also the view of the ancient Vedic texts and tradition.

#7120 VHS Video - 1 hour - List ~~$29.95~~ **$24.99** SAVE 17%

ORDER FORM

Other titles Gregg Braden references in his seminars and workshops.

Book

ITEM#	DESCRIPTION	PRICE
5640	Book of the Hopi	13.95
5645	The Book of Enoch	11.95
5637	Crystal & Dragon	19.95
5592	Emerald Tablets of Thoth-The-Atlantean	15.95
5591	Nothing In This Book is True	14.95
5594	Something In This Book is True	14.95
5601	Ancient Secret of the Flower of Life - Vol. 1	25.00
5602	Ancient Secret of the Flower of Life - Vol. 2	25.00
5630	Sacred Geometry	16.95
5635	Sacred Mirrors	29.95
5593	Nag Hammadi Library	21.00
5600	Essene Gospel of Peace (set of 4 books)	22.00

Video

ITEM#	DESCRIPTION	PRICE
7410	Mathematics for Lovers	20.00
7411	Cymatics III: Bringing Matter to Life	25.00
7412	Cymatics IV: Cymatic SoundScapes	25.00
7414	Cymatics: The Healing Nature of Sound	30.00
7590	Sacred Living Geometry (2-VHS)	34.95

Music

ITEM#	DESCRIPTION	PRICE
8050	Chaco Canyon - CD	16.95
8051	Chaco Canyon - Cassette	9.95
8100	Eternal Om - CD	16.95
8101	Eternal Om - Cassette	10.95
8060	Macchu Picchu Impressions - CD	16.95
8061	Macchu Picchu Impressions - Cassette	9.95
8111	Music for the Native Americans -Cassette	11.95
8110	Music for the Native Americans - CD	19.95
8065	Ocean Eclipse - CD	16.95
8070	Where the Earth Touches Stars - CD	16.95
8071	Where the Earth Touches Stars - Cassette	9.95
8130	Jungle of Joy - CD	16.95
8080	Higher Ground - CD	16.95
8081	Higher Ground - Cassette	9.95
8120	Beautiful World in Existence - CD	15.95
8121	Beautiful World in Existence - Cassette	10.95

SHIP TO:

PHONE:

ITEM#	DESCRIPTION	PRICE	QTY.	TOTAL

TOTAL =

SHIPPING/HANDLING =

SUBTOTAL =

WA Residents SALES TAX 8.8% =

GRAND TOTAL =

Credit Card Information

❑ Discover ❑ Visa

❑ MasterCard ❑ American Express

Card# ______________________

exp. date ______________________

name on card ______________________

signature ______________________

International Shipping Charges

PREPAID ORDERS: Due to the wide variety of products we offer, it is near impossible to pre-determine a fair international shipping charge for every situation. To ensure check customers are charged the correct cost of shipping, please e-mail (customerservice@radiobookstore.com), fax or call for the shipping charge. To pay by check, send an international bank money order payable in U.S. funds, drawn through a U.S. bank.

CREDIT CARD ORDERS: will be charged cost of shipping.

Videos are NTSC - American Format unless otherwise noted.

Shipping & Handling Charges

If the total falls between:	Add: US	Canada
0 - $9.99	$2.95	$ 5.95
10.00 - 24.99	$3.95	$ 7.95
25.00 - 49.99	$4.95	$ 9.95
50.00 - 74.99	$5.95	$12.95
75.00 - 99.99	$6.95	$14.95
Over $100.00	**Free**	$16.95

www.radiobookstore.com Phone 800-243-1438 (425) 455-1053 Fax (425) 455-1231